Violence in Place, Cultural and Environmental Wounding

Human life is intimately woven into place. Through nations and homelands, monuments and sacred sites it becomes the anchorage point for ethnic, cultural and national identities. Yet it is also place that becomes the battlefield, war zone, mass grave, desecrated site and destroyed landscape in the midst or aftermath of cultural wounding.

Much attention has been given to the impact of trauma and violence on human lives across generations, but what of the spaces in which it occurs? How does culturally prescribed violence impact upon place? And how do the non-human species with whom we coexist also suffer through episodes of conflict and violence? By identifying violence in place as a crisis of our times, and by encouraging both the witnessing and the diagnosing of harm, this book reveals the greater effects of cultural wounding. It problematises the habit of separating human life out from the ecologies in which it is held. If people and place are bound through kinship, whether through necessity and survival, or choice and abiding love, then wounding is co-terminus. The harms done to one will impact upon the other. Case studies from Australia, North and South America, Europe and the Pacific, illustrate the impact of violence in place, while supporting a campaign for methodologies that reveal the fullness of the relational bond between people and place.

The book will appeal to students and practitioners alike, with interests in cultural and human geography, anthropology, environmental humanities and moral ecology.

Amanda Kearney is Senior Lecturer in Anthropology at the University of New South Wales in Sydney, Australia.

Routledge Research in Culture, Space and Identity
Series editor: Dr. Jon Anderson, School of Planning and Geography, Cardiff University, UK

The *Routledge Research in Culture, Space and Identity Series* offers a forum for original and innovative research within cultural geography and connected fields. Titles within the series are empirically and theoretically informed and explore a range of dynamic and captivating topics. This series provides a forum for cutting edge research and new theoretical perspectives that reflect the wealth of research currently being undertaken. This series is aimed at upper-level undergraduates, research students and academics, appealing to geographers as well as the broader social sciences, arts and humanities.

For a full list of titles in this series, please visit
www.routledge.com/Routledge-Research-in-Culture-Space-and-Identity/
book-series/CSI

Memory, Place and Identity: Commemoration and remembrance of war and conflict
Edited by Danielle Drozdzewski, Sarah De Nardi and Emma Waterton

Surfing Spaces
Jon Anderson

Violence in Place, Cultural and Environmental Wounding
Amanda Kearney

Arts in Place: The arts, the urban and social practice
Cara Courage

Violence in Place, Cultural and Environmental Wounding

Amanda Kearney

LONDON AND NEW YORK

First published 2017
by Routledge
2 Park Square, Milton Park, Abingdon, Oxon OX14 4RN

and by Routledge
711 Third Avenue, New York, NY 10017

First issued in paperback 2018

Routledge is an imprint of the Taylor & Francis Group, an informa business

British Library Cataloguing in Publication Data
A catalogue record for this book is available from the British Library

Library of Congress Cataloging in Publication Data
A catalog record for this book has been requested

ISBN 13: 978-1-138-54646-2 (pbk)
ISBN 13: 978-1-138-92107-8 (hbk)

Typeset in Times New Roman
by Out of House Publishing

This book is dedicated to place and the co-presences with whom I share this world.

Contents

List of figures   ix
Acknowledgements   x

Introduction   1

PART I
Meeting place in the epistemological gap   13

1   Context and cultural wounding: the relational sphere of
    life in place   15

PART II
Witnessing place violence and the intent to harm   49

2   An ethnography of place harm   51

PART III
Diagnosing place harm   95

3   Destruction and designification   97

4   Social disorder: toponymic erasure and the making of
    harmful places   124

5   Elemental erasure and ecological decline   153

PART IV
Reinstating kinship and healing place   191

6   Kincentric ecology and seeking an axiological return   193

     Index   207

Figures

2.0   Yanyuwa Country, southwest Gulf of Carpentaria, Northern
Territory, Australia.   53

2.1   Approximate linguistic boundaries in the southwest Gulf
of Carpentaria, Northern Territory, Australia. The map on
the left shows pre-contact and early post-contact linguistic
boundaries, and on the right, present-day linguistic boundaries.   55

2.2   "They put the gates up everywhere now", gate at entrance to
Bing Bong Pastoral Station, 2003 (Photo source:
Amanda Kearney).   63

2.3   Bing Bong, Wurrulwiji in 2010, before the construction of a
large conveyer belt to transport iron ore for Western Desert
Resources (Photo source: Amanda Kearney).   64

2.4   Bing Bong, Wurrulwiji in 2014, after the introduction of
infrastructure to support iron ore transportation from the
Gulf of Carpentaria (Photo source: Amanda Kearney).   65

2.5   Close-up of operations at Bing Bong, Wurrulwiji, 2014
(Photo source: Amanda Kearney).   66

2.6   Signage, Bing Bong (Photo source: Amanda Kearney).   66

2.7   Signage, Bing Bong (Photo source: Amanda Kearney).   67

2.8   Aerial photograph of Bing Bong, highlighting the
designification efforts of agents of harm, across the wider area
(Photo source: Google Maps).   67

2.9   Bing Bong Port Facility, Glencore McArthur River Mine,
2015 (Photo source: Amanda Kearney).   80

2.10 Bing Bong Port Facility, Glencore McArthur River Mine,
2015 (Photo source: Amanda Kearney).   81

Acknowledgements

I set out to write a different kind of book, not like anything I'd written before. By that, I mean I wanted to write freely, within the conventions of academic and disciplinary traditions, but in a way that conveyed just how much I care about this topic. I wanted to write in a way that didn't diminish the knowledge and insights that Indigenous collaborators have shared with me. Their knowledge is neither secondary nor alternative, but rather is vital, and at the very centre of this work. This book is an attempt to reveal the knowledge gaps that have been produced by Western epistemologies and sustained through the effectiveness of a settler colonial state's epistemological violence. Across these pages, I strive to be with Indigenous knowledges in a more committed and decolonising way, ever-aware that I do this as a non-Indigenous person. I acknowledge my position as one that is firmly within the gap, yet I know that I am attempting to reach out for possibilities in the way others and I might come to know the world we inhabit. I sincerely hope to have done this with respect.

This book is a testimony to everything that Yanyuwa elders, middle and younger generations have taught me. In 2000, when I first arrived in Borroloola, an Indigenous community and township in northern Australia, I had no idea how much I could learn, nor how little I knew. I also did not anticipate just how kind people would be in teaching me to see the world in a new way. I was young and probably quite silly, but knew that if I sat and listened I might become a better kind of human. If I have in any way, I owe it all to Yanyuwa families, and the following incredible teachers: Dinah Norman Marrngawi, Annie Karrakayn, Jemima Miller Wuwarlu, Rosie Noble Wundirrimara, Mavis Timothy a-Muluwamara, Amy Friday Bajamalanya, Roddy Harvey Bayuma-Birribalanja, Dinny McDinny Nyilba, Pyro Dirdiyalma, Roddy Friday Mayalkarri, Gloria Friday, Graham Friday Mudaji, Leanne Norman Wulamara, Joanne Miller a-Yuluma and Rex Norman Wungunya.

If it were not for the kindness of my dear friend, mentor and fellow anthropologist John Bradley I would never have found my way to Borroloola, nor been so welcomed by Yanyuwa family members. John, you paved a way, one of deep ethics and openness to ways of knowing. For the past ten years we have returned time and again to Bing Bong, to walk the length of this important part of country. Every one of those pilgrimages is etched in my heart and mind, and for as long as I can I will return to that place in the hope that one day it may be lifted up again. My collaborations and friendships with Liam Brady, Steve Johnson and Philip Adgemis have been the scaffolding to support what I have attempted to do throughout this book. Our long conversations and long drives have provided clarity and something akin to confidence in this work. For that I thank each of you.

Gratitude is also due to Zora Simic, Kate Fullagar and Jessica Whyte who kindly read and commented on chapter drafts, provoking thoughts and discussions that were invaluable. I also wish to acknowledge the contributions of students who participated in my undergraduate courses 'People and Place' and 'Trauma and Violence' over the past few years at the University of New South Wales. You allowed me to use lectures as a time and space in which to work through my thoughts, and to explore possibilities and new pathways for configuring and valuing the place world. Those teaching moments have been essential to the germination and growth of this book.

To Kevin Kearney, my father, I say thank you for all of your support, belief and unfailing optimism. Knowing you are always there makes everything so much easier. Finally, to Marco Susino and our one-year escape to Tasmania. You and that time allowed this book to become a reality. We needed an escape, we found one, and from that, and the countless hours spent gardening and caring for my little place, I found my own true connect and became more aware and appreciative of sentiency in place. Between Tasmania and Borroloola, I feel so fortunate to have carved out an emotional geography of substance. Because of this, I can no longer accept the condition of unresponsive reflexivity. As hard as it might be to move reflexively and responsively through life, I know that it is because of Yanyuwa families and their country that I have become better equipped to do so.

Introduction

Place is a vital part of our human existence. It is the ground on which we stand, the home we defend, and the earth into which we might place the bodies of our deceased kin. In many respects it is also our kin, so central to humanity that without it we might be said not to exist. Despite this dependency and in spite of the intimacy that links people and place, there is a realisation that human action, as conflict, causes great harm to place via the destructions of war, the violence of ethnic tension, environmental strife and ecosystem degradation. Witnessing violence and the wounds that exist in place prompts reflection on the moral ecology of contemporary life; asking, what is it that we humans owe to the place world? This calls for reflection on what comes of place when it absorbs the violence of human conflict.

By identifying violence in place as a crisis of our times, and by encouraging both the witnessing and the diagnosing of harm, this book reveals the greater effects of cultural wounding. Coming to know the ways in which place absorbs and experiences human conflict problematises the habit of separating human life out from the ecologies in which it is held. If people and place are bound through kinship, whether through necessity and survival, or choice and abiding love, then wounding is co-terminus. The harms done to one will impact upon the other. By illustrating this point, and campaigning for methodologies that reveal the fullness of this relational bond, this book inches towards the possibility of living a kincentric ecology (inspired by Salmón 2000), or moral ecology, according to which, people and place are bound by a relational importance, which nests them together through the experience of harm as well as healing.

Commonly considered the context and milieu of life, place is a 'relational co-presence', envisioned as a vital shaping element in human life. Indeed, as the existing literature attests, place is many things: the physical environment and ecology, locale, homeland, ancestral landscape, and also a presence that exists in the mind, providing certainty and security. It is within and across such broad-ranging realities of place that human conflict occurs, and it is place, along with people, that bears the scars of violence and becomes the subject of trauma narratives. Understood as something that happens 'in place', and thus affects place, cultural wounding is a process of harm-doing that has wider implications beyond human suffering and hardship. Not wanting to diminish the magnitude of cultural wounding as a human experience, this book pivots on the principle that in order to better understand the implications of harm-doing and further a discourse and practice of healing and justice delivery for those who are culturally wounded, then the full extent of harms done must be explored and somehow better understood. Both the initial impact of wounding and its spreading effect into all aspects of human life, including place relations and place existence, deserve ongoing attention.

Cultural wounding – the violation of persons and their cultural lives through insult and injury, motivated by the desire to destroy or significantly harm this culture and its bearers – has greater effects than inducing harm on human life (Cook et al. 2003: 18; Kearney 2014). While it ruptures and assaults the cultural matrix that holds human groups together, in physical, emotional, spiritual and ideological senses, it also wreaks havoc on the place world. It does so through several mechanisms, including the destruction of environments and ecologies through warfare as a tactic designed to kill or harm human life, and the social disordering of place through erasure of place order and toponymic distinctiveness, as orchestrated efforts to devalue and compromise cultural presences in space and time. The capacity to deliver harm in place as a result of human conflict and violence against ethnic or cultural others is traced to the nexus in which place and human life coexist. This is something more than place as context or incidental site of conflict, but instead positions place as a co-presence, in that where human life exists and is rooted into the world are found places of meaning and worth. In the process of destroying or deeply harming a cultural group or those who declare an ethnic distinction, attacks on the most necessary and meaningful aspects of life are launched. Place becomes the object of hateful desires.

Place harm as one expression of a greater motive to culturally wound, is achieved through deliberate and sanctioned acts such as colonisation (neo-colonialism), war, genocide and ethnic cleansing. So too is it brought about through epidemics and destruction of environments, taking of lands, forced removal and enslavement. The effects of violence are measured by the prevalence of emotional anguish, physical suffering, erasure and destruction. Such tragic events can be engaged in a moment or enduringly encountered by an individual, a family, a place, community of non-human species or entire ecosystem. This is the experiential nature of violence and trauma (Eyerman et al. 2011; Alexander et al. 2004). The greater effects of cultural wounding, as they come to impact on place, are best understood through interrelation and interconnection of distinct, irreducible and interrelated components; people, place and place elements. Cultural wounding often works to close off what is conceived of as cultural space and seeks to harm the possible futures for the people and places that receive its violence.

Through wounding events, there can emerge a sequence of actors: both human and non-human. These include the perpetrators of violence, recipients of harm and those who witness the events and their calamitous fallout. In the case of cultural wounding, the agent of harm is overwhelmingly human, not just any human but more specifically those invested with power and influence, capable of an epistemological, ontological and axiological habit that disengages morally from a relationality that positions people as valuable and also place as valuable. This human is what Erikson (1994) describes as a "new species of trouble". Destructions and violence delivered by this agent of harm, are, he writes, "[d]ischarged along the lightning conductors that protect power, preferentially affect[ing] the underprivileged: third world inhabitants; ethnic minorities; physical and manual workers; women, or isolated old people". Such agents of harm act with intent and often harbour an unwillingness to care. In doing so, their actions frequently efface the place world, as a direct blow designed to crush ethnic solidarity, cultural vitality and the human spirit, while inducing fear and uncertainty about where people might anchor their lives and live safely. The recipients of harm and those who witness it are both human and non-human, a situation that is explored more fully in Chapter 1.

Throughout this book, I invoke perspectivism and hyper-relativism, guided by the principle that the world is populated by a vast number of 'agents' or co-presences, all of which participate in and actuate a sentient universe. This relies upon kinship, and resisting the practice of separating human life out from the contexts and sites of meaning in which it operates, while acknowledging that the relationships formed between people and place occur within particular epistemic spaces. These spaces are in turn governed by ethics and axiologies. The relationships and ethics that guide interactions between people and place are expressed by ontologies that convey cultural distinctions in how people might conceive of place, how place forms and maintains itself, and how human life interacts with and is influenced by place. So vast is the field of potentials, yet so peripheral has been academic attention to the possibility of place being an agent unto itself. Many Indigenous epistemologies rely upon a distinguishing 'perspectival quality', a conception according to which the world is inhabited by different sorts of subjects or persons. This epistemology of place and its associated ontologies and axiologies are what guide this book's approach to place and conceptualisation of place harm. The latter made possible because according to an Indigenous epistemology, there is both sentiency and agency in place. The importance of this approach is traced to the strength it offers in redressing the disconnect between people and place commonly encountered in Western epistemologies and neoliberalism. The disconnect manifests in a failure to care or a denial of kinship as that which ensures responsibility and mutuality.

Axiological crises originate primarily from cultural causes, as conflicting social interests or an assumed incompatibility of two or more cultural systems, and may be configured as internal or relational. In the first instance, the crisis may refer to transformations in the form or systems of values within a given culture, configured as dilemmas as to what kind of life is worth living, what kind of life is worth preserving and uncertainty or dissent over 'how to be'. Relational crises, on the other hand, are located within the gaps that form between axiologies across or between cultures. Furthermore, they may be the effects of axiological ignorance on the behalf of a dominant culture whereby cynicism and/or unrestrained disregard of the ideals, rights and interests of other peoples soon becomes intolerable and inevitably leads to open conflict and 'survival crises'. It is relational crises that inform a discussion of place harm as a greater effect of cultural wounding. And it is these that express themselves through the attitudes (epistemologies) and actions (ontologies) that form around place and its disordering, such as a failure to care or prevent harm, the narrating of harm as consequential and unavoidable, loss as the cost of human development, and principles of 'time to move on'.

It is hyper-relativism that leads me to this methodological leaning, as both anthropologist and cultural geographer. Hewitt helps to orients us towards the direction of place as co-presence, when he reminds the reader:

> Just as biological life may be called a set of activities intended to resist death, so our place and world are at least partly a means to resist psychosocial and cultural dissolution. That becomes more readily apparent when war or other calamities damage and threaten to destroy land and settlements. Unfortunately, war also mobilizes the highly charged and dangerous dialectic of place attachment: the perceived antithesis of 'our' places or homeland and 'theirs'. Sustained in latent if not overt forms in peacetime, this polarization has produced unbridled sentimentalizing of one's own while dehumanizing the enemy's people and land. That seems an essential step in cultivating the readiness to destroy the latter and bear with progressive devastation at home.
>
> (Hewitt 1983: 258)

Contained in Hewitt's writing are allusions to the axiology that supports place harm; a cognitive framing that enables war on place, waged in the name of attacking human life, with the intent to cause death and destruction. Axiology is treated here as the perception of worth and sense of value as it is generated within any cultural context. It relates to what people come to value and how they determine worth as culturally prescribed, thus there are distinctions to be found in the way that people perceive and prioritise aspects of life.[1] Writing specifically about area bombardment and the fate of urban places, he reflects on the ontics of place annihilation and the ethics that support the occurrence of violence and that linger in the aftermath. Hewitt (1983: 258) cautions that places, like biological life, participate in the "distinctive features of intelligent life, its creativity and search for order" thus rendering place a highly effective target when the intent is to harm human life. Annihilation of place is tantamount to erasure of foundations and instating of chaos, for place is one part of the creative expression of life and meaning, revealing the relational universe in which human life exists and holds strong.

Places, he provokes, also share the problems of survival and mortality as do humans through their biological existence. This defines place as an essential co-presence that insures against mortality by providing shelter and the anchorage point for identities and bodies. So too the very existence of place becomes a testament to human survival. If place prevails then the likelihood of human life achieving the same measure of survival is greatly increased. This aspect of Hewitt's work might also be read as a further caution, reminding the reader that place itself is vulnerable when faced with threats of violence. It too has an essence that might be expunged when mortally wounded and a sentiency that may grapple to survive when made the subject of violence through war and destruction. The possibility of this is explained further:

> to eradicate or extirpate, is not directly to destroy plant, animal or person as individual detached organisms, but to uproot them. To exterminate is literally to kill by geography, not necessarily damaging an organism, but driving it beyond the bounds. Exiles, expellees, and others compulsorily displaced have often described their plight as tantamount to cultural starvation, if not death or worse. That may seem too subjective to situate the problem squarely within modern geographic scholarship. Nevertheless, the geographical impact of the uprooting and removal of tens of millions of people from their longtime homes in the wars, expulsions, and evacuations of this century cannot but be enormous.
>
> (Hewitt 1983: 259)

The geographical impacts and uprooting effects of war and conflict to which Hewitt (1983) refers are here treated as violence against place. They are equal in their effect on human life and place agents. They cause harm and are, in instances of cultural wounding, enacted with intent. That said, one of the difficulties faced when writing of violence, of any kind, is that, "like power, violence is essentially contested: everyone knows it exists, but no one agrees on what actually constitutes the phenomenon" (Robben and Nordstrom 1995: 5). "Vested interests, personal history, ideological loyalties, propaganda, and a dearth of firsthand information ensure that many 'definitions' of violence are powerful fictions and negotiated half-truths" (Robben and Nordstrom 1995: 5). "Most of the time, people are attending to the routine tasks of their lives, to eating, dressing, bathing, working, and conversing. Conceiving of violence as a dimension of living rather than as a domain of death obliges researchers to study violence within the immediacy of its manifestation" (Robben and Nordstrom 1995: 6). "Violence is also an intricately layered phenomenon. Each participant, each witness to violence, brings his or her own perspective. These testimonies can vary dramatically" (Robben and Nordstrom 1995: 5).

Accounts of violence are exercises in making sense of the incomprehensible, invariably delivered as narratives or testimonies of cultural trauma, cultural wounding and survival (see Alexander 2004; Alexander et al. 2004). It is the carrier group who lives through and recalls violence and trauma, and that shares such testimonies with a constellation of witnesses. Testimonies may take the form of speech acts, pre-lingual expressions and other embodiments of pain and hardship (such as bodily habits, tattoos, dispositions) (Connerton 2011). So too geography and architecture may become testimonials to what has occurred. Place enters the frame here as a carrier itself, and also as a witness to violence and trauma. It is on the one hand capable of absorbing the effects of cultural wounding and rendering this through expressions of a physical and tangible kind, while also being capable of holding onto the effects of violence through intangible expressions of disorder, such as spectral traces, absence and silence. Even the most horrific acts of aggression do not stand as isolated exemplars of a 'thing' called violence but cast ripples that reconfigure lives and the place world in the most dramatic of ways, affecting constructs of identity in the present, potentialities for the future and even renditions of the past (Robben and Nordstrom 1995: 5).

Despite these possibilities and the expansiveness of harm-doing, less common are accounts of violence and testimonies of cultural wounding that go beyond speech acts and orality as their communicative modes. Examples of such may be the expressions of health and ill-health, order and disorder among populations of non-human animals, amid places and features on the landscape, and as articulated through the balance or imbalance of ecologies as key elements of the place world. Why is it that such accounts of witnessing and experiencing violence are less readily attended to or that such occurrences are less often considered acts of violence? There is evidence of violent assault as axiological disregard, overt destruction and enduring toxicity in so many places, such as the burning oil fields of Kuwait, among homeless dog populations in Sarajevo, at mass grave sites in Bosnia or where the parched landscapes of colonial frontiers are riddled with feral species and devoid of Indigenous life. These are direct evidence of violent acts, culturally constructed and delivered via the ontics of ethnic conflict, colonialism and war.

Overlooking the saturating effects of violence, as manifest in such examples, means neglecting to attend to the obvious signs of harm having been done to place and the dismissal of any relational standing between people and place. Violence has become a dimension of people and place's existence, not something external to society and culture that 'happens' but a pervasive presence in a vast number of places (Robben and Nordstrom 1995: 2). Irrespective of its form, what distinguishes violence in place, as a happening, is the act of denial. As an action, or social construction, violence requires a form of committed axiological retreat, which supports the denial of value, relevancy, order and mentionable consequence. Axiological retreat invokes principles of disregard and moral disengagement, at a level so profoundly normalised that the question of care passes into oblivion.

About this book

Place is important to human life, hence it is claimed, fought over and capable of being harmed. Why does this matter? Some would say that place no longer matters in an increasingly globalised and disconnected world, or that highly mobile ethnic groups are capable of 'making place' wherever a 'sense of community' is found to exist. Malpas (2009) reflects that "there are many contemporary theorists who would argue that, if the advent of globalization does imply a loss of any real sense of place, then this is no bad thing and that the sooner we can discard the idea of a special connection to place, the better". This book veers away from such a view, instead committing to the view that rectifying humanity in the midst of prevailing ethnic conflict, violence and trauma, and rapid (human-induced) environmental and ecological decline requires a return to the intimate kinship that exists between people and place. This relational reminder may provide a vital step in redressing the harms done, by more fully conceiving of the universe of interactions that are implicated both in the experience of cultural wounding, and the process of healing. There is a necessary character to place, a particularity to each place that means its importance both as a spatial derivative, an ecology, and expression of order and meaning, cannot be dismissed.

Ultimately, this book aims to establish cultural wounding as complex, multi-sited and relational, in which case the experiences of people cannot be disassociated from place. Sites of violence and cultural wounding come to exist in landscapes scarred by aerial bombing, neighbourhoods reduced to rubble, ethnically cleansed villages, death camps or landscapes denuded of trees and littered with dying livestock. What is often most confronting about these places is an overwhelming absence of life and sense of profound disorder. The people might be gone or have been replaced by another ethnic group, species may have disappeared and ecosystems fallen into toxic disarray, yet place often holds on, harbouring the pain, while revealing another kind of story about conflict and violence. As dark tourism and studies of traumascapes reveal, people flock to visit locations such as these as part of a growing curiosity or need to witness and reconnect with place (see Tumarkin 2005). They might attend to place as a memorial and defiant gesture to the structural amnesia that insists they forget, or might pass through place with a vague knowledge of events and still find themselves overcome by a sensory and emotional geography that hints at past harms. Whatever motivates the interaction with a wounded space as a place harmed, as a return or remembrance, the impact of violence is often powerfully communicated, not the least because these places highlight the encompassing nature of ethnic conflict and cultural wounding, but because they are a testimony to human relations and important, often incomprehensible moments in time. What is left in place also defies the notion that suffering is exclusive to human beings. Hauntings, scars and trauma may linger, thus suggesting something of what place experiences as a result of violence and conflict. Signalling aspects of what has occurred, these traces reiterate that the effects of wounding cannot always be shaken or ignored, whether they are fully understood or not.

Living with or healing from the greater effects of cultural wounding, as I have revealed through earlier work (Kearney 2014), is a formidable task that involves more than treating bodies, remembering atrocities, or legal and political redress. While these are vital and necessary steps towards restitution and justice, healing and reconciling ethnic conflict and its resulting cultural wounding requires interacting with the ecology of wounding. If ethnic violence and conflict is viewed in terms of an ecological approach, whereby human conflict is part of a wider matrix of relations, constituted by the world (as made up of physical, spiritual, political and social universes) in which we live, then the human experience cannot be disentangled from that of place and other constitutive elements of place, such as non-human animals and ancestral beings. If this possibility is explored then it paves the way for a discussion of how harm directed at people on the basis of perceived ethnic or cultural difference or an associated axiological disconnect, is absorbed into place and therefore able to hold on or infuse future relations and interactions between people and also places. How its effects can be shifted requires knowing the depth of the impact.

Part I of this book offers the reader a rationale for this work. It sets the scene for an analysis pivoting around cultural wounding, which positions humans, place and non-human species in a profound relationship where violence committed against one have repercussions on the other. Raising the possibility of agency and sentiency across the entire scope of human/place/non-human agents, it resolves to make central the view that humans are not the only agent capable of authoring place and determining the nature and value of place. Meaning-making that adheres to and expresses inherent place order is undertaken by all place elements, as is visible in the actions of non-human animals, flora, ancestral beings, geological formations and other co-presences that make up the place world. A relational logic orients the perceptual field towards inclusion of all place elements, which means that if humans suffer amid war and violence, then place elements do as well.

Having distinguished place as a recipient of harm, brought on by the greater effects of cultural wounding, a conceptual framework for examining place as a relational other to human life is necessary. This framework must be extensive in its scope, liberated from the anxieties of binary thinking (nature/ culture, human/non-human, scientific/spiritual), and capable of conceiving place as a meaningful co-presence. Chapter 1 outlines the epistemological gaps between Western and Indigenous pedagogies of place that have prevented such thinking. Drawing together Indigenous epistemologies of place and elements of a contemporary social and cultural geography is one way of committing to first 'knowing the gap', then 'sitting in the gap' and perhaps even finding a join. This methodological intervention engages with the epistemologies and ontologies that allow us to fully witness place harm, namely Indigenous pedagogies of place, decolonising methodologies, kincentric and nested ecologies and emotional geography. The concepts adopted throughout Chapter 1 and throughout the entire work, also take their inspiration from wider studies on cultural wounding (Kearney 2014), memory and mourning (Connerton 2011), cultural trauma (Alexander 2011; Alexander et al. 2004), along with Rose's (2004, 2011) work on wounded space. Erikson's (1976, 1994) work on place destruction, community loss and the human as a "new species of trouble" also provide important foundations for the forthcoming conceptualisations of place harm and survival.

Part II of the book encourages and practices the act of witnessing. This means sitting with the violence of place harm, observing the destruction and more closely coming to appreciate the experience of wounding for place and its environs. This takes the focus of discussion to a sequence of events that have led to the wounding of people and place in an Indigenous community in northern Australia. By recounting the genealogy of a place attack and potential death, as situated on Indigenous Australian homelands, we are compelled to more fully appreciate place harm and the effects of violence on place. Moving through Chapter 2, the reader is positioned as secondary witness to these events and through the telling of place harm as recounted by Indigenous place kin, in this case members of the Yanyuwa Indigenous community, is encouraged to engage with decolonising principles in the reading and interpretation of harm in place. While it is not possible to engage the testimony of place itself (due to the limits of text-based speech acts), it is hoped that by reflecting on 15 years of collaborative ethnography, and sticking as closely as possible to the testimonies of Yanyuwa place kin, there is found a link into place consciousness and being, in accordance with Yanyuwa epistemology. Yanyuwa understand place as saturated with the presences of the ancestors and creation beings that generated an entire world of meaning in the southwest Gulf of Carpentaria, northern Australia. That the two (people and place), commune through a kincentric ecology is a key part of this Indigenous community's epistemology, ontology and axiology.

In Part III of the book, the emphasis is placed on diagnosing place harm. This begins with a critical rethink on what might constitute an agent of harm. How is the human agent of harm distinct in its ideologies and praxis from other threats, namely those of 'natural' disasters? The methods by which people harm place, with the intent to destroy ethnic and other cultural presences, are subsequently distinguished through a discourse of 'destruction and designification', 'social disorder' through renaming, the erasure of toponymic distinctiveness and the creation of sites of suffering, and lastly as 'elemental erasure and ecological decline', by way of toxicity, contamination and killing of non-human life. By coming to know the patterns of place harm, it is possible to recognise its presence in contexts all over the world. The reader is introduced to these major forms of place harm throughout Chapters 3, 4 and 5, as effort is dedicated to diagnosing the ways in which place might be harmed, drawing on historical and contemporary instances of place harm across a vast number of countries and regions. Throughout these chapters, the wounding of place is exposed for its methods and motivations. By focusing on the suffering experienced by place and its non-human counterparts in the midst of cultural conflict, violence and axiological retreat, the relational experience that is cultural wounding is rendered explicit for the reader.

In the final section of the book, Part IV, the discussion is directed towards consideration of the important questions, can places heal? How might they heal? Seeking pathways to an axiological return, in which kinship between people and place is found, and inspired, is the final objective here. This requires a reflective discussion on the principles and relational commitments essential to kincentric ecology, as a heuristic and practical device aimed at redressing place harms that occur worldwide. Most importantly the challenge of how these principles can be lived and re-entered into the normative practice of everyday place interactions among cultural and ethnic groups worldwide, as inclusive of Indigenous and non-Indigenous identities, is addressed. In concluding, it is proposed that kincentric ecology provides the most substantial and proven epistemology for realising ontological and axiological shifts in human conceptualisations and relationships with place.

Note

1 Axiology is based on the principle that humans will and can regulate the way that various objects, situations or events are perceived, to the benefit of survival and of quality of life. These values can be configured in many ways – economic, moral, or aesthetic – and value may be relative or absolute, inherent or diffuse. It is impossible to overstate the importance of axiology in determining behaviour and views, for it is the foundation for conscious judgements and decisions, and therefore the basis for purposive thought and action. Although some acts may be reflexive or instinctive and cannot therefore be ascribed to conscious beliefs about value, any action based on even the most cursory reflection has its foundation in standards of what is good or bad, right or wrong.

References

Alexander, J. 2004. Toward a Theory of Cultural Trauma, in *Cultural Trauma and Collective Identity*. Edited by J. Alexander, R. Eyerman, B. Giesen, N. Smelser and P. Sztompka, pp. 1–30. Berkeley, California: University of California Press.

Alexander, J. 2011. *Trauma: A Social Theory*. Cambridge: Polity Press.

Alexander, J., Eyerman, R., Giesen, B., Smelser, N. and Sztompka, P. (eds). 2004. *Cultural Trauma and Collective Identity*. Berkeley, California: University of California Press.

Connerton, P. 2011. *The Spirit of Mourning: History, Memory and the Body*. Cambridge: Cambridge University Press.

Cook, B., Withy, K. and Tarallo-Jensen, L. 2003. Cultural Trauma, Hawaiian Spirituality and Contemporary Health Status. *California Journal of Health* Vol. 1: 10–24.

Erikson. K. 1976. *Everything in its Path: Destruction of Community in the Buffalo Creek Flood*. New York and London: Simon and Schuster Paperbacks.

Erikson, K. 1994. *A New Species of Trouble: Explorations in Disaster, Trauma and Community*. New York: W.W. Norton and Co.

Eyerman, R., Alexander, J. and Butler Breese, E. (eds.). 2011. *Narrating Trauma: On the Impact of Collective Suffering*. Boulder, Colorado: Paradigm.

Hewitt, K. 1983. Place Annihilation: Area Bombing and the Fate of Urban Places. *Annals of the Association of American Geographers* Vol. 73(2): 257–284.

Kearney, A. 2014. *Cultural Wounding, Healing and Emerging Ethnicities: What Happens When the Wounded Survive?* New York: Palgrave MacMillan.

Malpas, J. 2009. Place and Human Being. *Environmental and Architectural Phenomenology Newsletter* Vol. 20(3): 19–23.

Robben, A. and Nordstrom, C. 1995. Introduction, in *Fieldwork Under Fire: Contemporary Studies of Violence and Survival*. Edited by A. Robben and C. Nordstrom, pp. 1–23. Berkeley, California: University of California Press.

Rose, D.B. 2004. *Reports from a Wild Country: Ethics for Decolonisation*. Sydney: UNSW Press.

Rose, D.B. 2011. *Wild Dog Dreaming: Love and Extinction*. Charlottesville and London: University of Virginia Press.

Salmón, E. 2000. Kincentric Ecology: Indigenous Perceptions of the Human–Nature Relationship. *Ecological Applications* Vol. 10(5): 1327–1332.

Tumarkin, M. 2005. *Traumascapes: The Power and Fate of Places Transformed by Tragedy*. Carlton, Victoria: University of Melbourne Press.

Part I

Meeting place in the epistemological gap

This book cautions against any tendency to imagine place as the backdrop to human life, or as a stage on which conflict and ethnic violence act out their ugly drama. Place is not an inert field into which human aggressions and the violence of ethnic conflict are projected, rather it is a fully absorbing and responsive co-presence that witnesses and experiences these acts as part of its own biography and existence. To fully realise this statement, Part I of this book is tasked with finding a methodology capable of presenting place as an agent, sentient co-presence and kin to human life. It is one part of a substantive relationship that justifies and delimits human existence and experience. It is not enough to simply state that place feels and lives through human conflict, for such arguments lie vulnerable to scepticism or charges of romanticism. What is needed is a fully operational methodology that epistemologically grasps at place-meaning from a new direction, drawing together ontological and axiological principles from across cultures to render an understanding of place as fully sentient and capable of response and action.

To begin, Chapter 1 introduces place as one part of the relational sphere of life. It works to distinguish place as an agent capable of being harmed but also as capable of becoming an instrument of harm when reinscribed with strange and often violent meaning by those who co-opt it into a wounding agenda. The methodologies that progress this vision of place are Indigenous epistemologies of place and phenomenological geography, the former engaged most thoughtfully and with decolonising principles in mind. Together these allow for a fuller witnessing of place and place harm, and when combined with the principles of nesting, emotional geography and kincentric ecology, the relational sphere of life in place and place in life is brought to light.

1 Context and cultural wounding
The relational sphere of life in place

Accounts of human conflict, violence and the quest for survival are never far away in the era of the 24-hour news cycle. At any moment we are drawn in to witness these struggles. As one part of the motive for violence, ethnic difference and subsequent hatred fuel a vast number of conflict situations. Conflicts, many historically motivated, erupt over disputed territories; land, sea and resource rights; political and ideological discord. Seen operating in public are tensions that seep into the private realms of personal lives, making their presence felt within familial environments, homes and villages, townships and local communities. We see ethnic conflict and violence occurring in the context of the state, across international borders, at the regional and local level and most intimately in the hearts of those who suffer abuses.

Violence, trauma and the cultural wounding that results have considerable spatial presence; so far-reaching that as they appear on our television screens, we are jettisoned for a moment into the places of conflict and into the often extreme everyday lives of those who experience ethnic prejudice and the aggressions that lead to cultural wounding. Although we may turn away, a witnessing has occurred. As we watch journalists scramble over rubble, or ethnic minorities march the streets in defiance of the state, and as asylum-seekers board unseaworthy vessels fleeing one place in search of another, cultural wounding threatens and takes hold. Once it takes hold, it is etched onto occupied territories and the bodies of displaced peoples, expressed as the loss of home and the destruction of environments once capable of sustaining human life. Dispossession of Indigenous homelands for state-sponsored developments and the commemoration of past violence through the construction of gleaming memorials also mark this scattered landscape of wounded places through which we often tentatively move.

Shown to us in the hope of eliciting an empathic response, the reality is that for many, accounts of conflict and suffering are abstract and by virtue of birth or good fortune are removed from the experience of everyday life. Yet we are called upon to witness these events and therefore participate in what is an 'ecology of cultural wounding and healing', which positions some as witness, some as the wounded and others as the perpetrators of violence. This book, written from the perspective of a witness, seeks to understand trauma and violence 'in place', in the hope of generating a deeper understanding of interethnic violence, cultural wounding and the impact these have not only on human life but also place, and non-human life as part of a network of co-presences. As an ethnographer, I have listened for over a decade and a half to accounts of violence that led to cultural wounding, as told by Indigenous Australians and African descendants in Brazil. Many times I've struggled to understand the magnitude of these experiences, yet when told to me 'in place', at the site of the experience, or when left to walk amid the remains of place, I believe I have inched closer to appreciating the extent of violence or at least its emotional toll. By visiting Indigenous homelands and seeing what has been taken, lost or destroyed and by standing in the ruins of slave ports in Brazil, I have tried to feel 'in place' the violence that has occurred by way of colonisation and slavery. Augmenting accounts of violence, trauma and cultural wounding with the sensory and emotional geographies of place has raised many questions for me, about the ways in which place bears the scars of conflict and intentional harm, and also how place itself might actually experience violence and the cultural wounding of its human counterpart. Exploring these concerns requires that I examine people and place relations, emphasising kinship and mutual constitution as the substance that joins the two.

Based on the principle of relatedness, the concepts and ethnographic case studies developed throughout this book aim to articulate an ecology of wounding that, once recognised, might elicit greater empathy for the effects of violence and wounding not only on human life, but also on place and on the non-human life with which we coexist – and that in some contexts are vividly cast as our kin. I do this by adopting a methodology that combines decolonising principles and Indigenous epistemologies of place and place agency. These are what allow me to more fully construct a sense of nested and kincentric ecology. A methodology of this kind renders possible a sense of kinship between people and place. Place is defined in such a way as to recognise its agency and responsiveness to human presences. It is hoped that by examining the relational sphere of life in place, an appreciation for place agency and experiential capacity can be reached; one that matches the human capacity to be harmed and to suffer. By locating people and place along an equal register of vulnerability in moments of cultural wounding and violence, these experiences are described as shared and mutually lived through.

Meeting violence and wounding in place

Rose (2004: 34) describes wounded space as that which "has been torn and fractured by violence and exile … pitted with sites where life has been irretrievably killed". Wounding is mapped in place by the "legacy of genocide and ecocide, savagery that has ripped at the earth and at the lives of many who inhabit it, a conquest over lands that has supported an agenda of devastation so sudden and massive that we might never fully grasp the consequences" (Rose 2004: 35). Practitioners in place studies, human and cultural geography and anthropology have sought ways to convey the magnitude of violence in place, expressed through dark places, urbicide, topocide and domocide, place annihilation and erasure (Abujidi 2014; Falah 1996; Porteous and Smith 2001; Sather-Wagstaff 2011; Trigg 2009; Tumarkin 2005). Research into cultural trauma and cultural wounding, as well as place designification and resignification has also honed a reflexive awareness for what happens in place when human conflict occurs, predicated on ethnic difference, perceived marginality and the alleged undesirability of some bodies over others (Alexander et al. 2004; Alexander 2004a, 2004b, 2011; Bracken 2002; Eyerman et al. 2011; Fassin and Rechtman 2009).

The loss or destruction of place resonates powerfully in the accounts of those affected by forms of forced removal and constitutive cultural shift brought on by ethnic conflict. Decisive and violent action befalls certain ethnic and cultural groups often on the grounds of place contest and the desirability of the land, sea and its resources. Livelihoods and ways of being are deemed 'out of place' or impediments to the forceful taking of territories. Subsequently they are positioned in the way of harm. Place, often a highly desired commodity in interethnic contact zones, is overwhelmingly pressed upon to provide refuge for some ethnic groups, while becoming a site for the banishment and erasure of others. The years are marked by and remembered for these events, or the denial of their having occurred. People who live through such experiences and the places that come to harbour these events are imprinted with the legacy of violence. The overwhelming shock for witnesses to this violence, whether immediately experienced or witnessed from afar, comes from the loss of human life, and the depravity of what humans will do to one another under the guise of conflict, and even ambivalence. On this human compulsion to wound and harm I have written, utilising the themes of cultural trauma, cultural wounding and motivated healing (Kearney 2014). These themes have emerged out of my work as an anthropologist and have been examined through collaborations with Indigenous families in Australia and African descendant and black rights groups in Brazil.

In both Australia and Brazil, people have experienced cultural wounding (Kearney 2014: 3).[1] For Indigenous ethnic groups in Australia and African descendants in Brazil, their histories and present realities are marked by loss and hardship on the basis of an alleged otherness and imposed marginality by powerful ethnic majorities. So too have both groups faced a profound and forced reconfiguration of the place relations that make them who they are. Whether through theft of place or removal from place, the relational world these ethnic groups occupy has been fundamentally rearranged. Theft, destruction and designification characterise just some of the deeply wounding strategies used against people and place in the context of Indigenous homelands. For African descendants, their wounding has also mapped onto place, leaving its signature in the very separation of people and place and construction of new, deeply disturbing places, such as those created for containment and captivity. The erasure of home or breaking of kinship with home through forced removal left African nations, tribal homelands and local communities without the men and women that constituted them. While place has come to bear the scars of cultural wounding, or has emerged out of the human suffering that slavery brought, it also figures as centrally important in healing projects across both ethnographic contexts. Place is often scripted as a refuge, offering security and the surety of ethnic belonging that is so longed for in the healing journey. Even those sites of great sadness and horrific memory can be reclaimed as essential to present healing and remembrance of hardship. It is as if lines of kinship offer the most secure pathway to healing and recuperation. Thus it is through a denial of kinship that wounding is made possible and may be sustained over generations. Instances of cultural wounding in both cases map heavily onto place. The most obvious incorporation of place into trauma narratives takes the form of recounting the injustice of place theft and destruction; the embodied suffering that comes with removal from place, and a deep longing to return. So too the paradox of return looms large in the narratives of those who have been wounded, often left asking, to what do I return? How do I reinstate kinship when my place kin is gone?

Returning home or seeking restitution of territories to which one belongs have become cornerstones of transitional justice efforts and reconciliation agendas in Australia and Brazil, and refuge has been found in places, whether home or not, that are capable of holding and nourishing ethnic constituents as a community of kin. For Indigenous Australians, place is often spoken of as longing for its people, yearning the very presence of its human kin. Rose (1996: 7) writes of Indigenous homelands, or country as a "nourishing terrain", "a place that gives and receives life. Not just imagined or represented, it is lived in and lived with." Among African Brazilian rights groups and African-Brazilian communities, place remains a nourishing terrain despite generations of loss and hardship. This has emerged through a powerful rhetoric of return and the portability of place. For quilombolas, there is the desire to legitimate land holdings that come down the generations from those who escaped enslavement and sought a 'future in place' for African Brazilians. For members of the African Diaspora, place lies both in the transatlantic voyage, as it does in African nations of origin, and present-day black communities throughout Brazil. Mama Africa is the largest and most looming of place concepts, both conceived of as myth, body, woman, kin and embodiment of journeying, and capable of taking in all across the Diaspora. Mama Africa becomes a site of meaning, a movable force of place for those who cannot return or find the pathways back or for whom their identity as African descendant is now powerfully enmeshed with black Brazilian identity. Mama Africa can also be found distributed across continents, a vast network of places thus ensuring her presence and survival. She is the most pervasive of places, both real and imagined. Reflecting on ethnographic insights drawn from my time in Australian and Brazilian contexts, what is striking is the extent to which the people and place relationship deserves a deeper reading in contexts of violence and wounding.

After 15 years of working with those who are structurally and historically oppressed through encounters with cultural wounding, I have become acutely aware of the greater effects of violence. My mind has turned to the places or contexts in which experiences of wounding 'took place' in the past and continue to 'take place' today. I note that with many of the trauma narratives I have recorded, events are traced to particular locations, places, homelands and even bodies of water. For Indigenous Australians, the very experience of cultural wounding cannot be disconnected from the colonial project's desire to possess place and resignify it under the name of the British Crown. Place and possession of it stands often as the original motive for violence and wounding. For African Brazilians who recall ancestral narratives of enslavement and hardship, place echoes as the social and cultural universe from which ancestors were first taken, and the cruel world into which they were thrust remade as slaves. Via the transatlantic journey taken by the millions of African persons enslaved, places that had no prior connection became bridged and thus inextricably linked for generations. The deep complexities in how descendants perceive and engage with the place from whence their ancestors came and those places they journeyed through can never be underestimated, even in Brazil today.

Methodology

Taking a lead from Indigenous epistemologies, along with phenomenological geography, and the contributions of Tuan (1974a, 1974b, 1977), Relph (1976, 1985), Seamon (1996, 2012; Seamon and Sowers 2008) and Buttimer (1976), much of my existing work on place has been drawn to and from the experiences of the everyday. From the lived encounter of an emotional geography, Indigenous place-naming, and the everyday violence of deep colonising, a phenomenological approach has highlighted the "multivariant structure" of place as a sophisticated and complex existential presence in our life, as well as spirited and geographical ensemble saturated with meaning and potency (Seamon 2012: 3). "A phenomenological approach 'offers a way to deal with the "richness" of place, where the ecological and the cultural, the human and non-human, the local and the global, and the real and the imaginary all become bound together in particular formations in particular places'" (Jones and Cloke 2002: 6, in Johnson 2012: 835). Concerned with the greater effects of cultural wounding, this book needs at its disposal a methodology that supports an exploration of place as relationally constituted by human life. This comes from a phenomenological approach to perception and existence, which contextualises interethnic conflict *in situ*. People and place are intimately linked in the moment of harm, and they remain so through the experience of trauma, enduring hardship and healing.

Establishing kinship between these two agents, people and place are positioned as conscious presences through their respective agency and sentiency (see Plumwood 2002; Rose 2013a). In which case, both are perceptual subjects, with each other as the perceptual object. The process of enmeshing one's self with place is a phenomenological one through which the perceptual being brings to fruition an experience of place relative to conscious and unconscious structures of meaning (Bradley 2001; Kearney 2009a, 2009b; Merlan 1998). Kinship, as a form of relatedness and mutual constitution, acts as a conductor through which the relational presence of people and place is brought to life. Whether framed as an intimate other, the tangible expression of one's ancestry or that which is longed for as a distant home long since departed, kinship is one way of defining the bond between people and place. In its most basic form, kinship refers to a set of socially determined relations that are "predicated upon cultural conceptions that specify the processes by which an individual comes into being and develops into a complete social person [entity]" (Kelly 1993: 521–522). Place is an essential part of this social movement. Within any cultural context, it is what mediates, constrains, delimits or even allows performance and celebration of certain ways of being. By constituting and reinforcing particular human presences or by being the site of human life, place renders people distinct. A homeland can justify a people or a journey through place might distinguish a cultural narrative. In turn, place can determine the extent to which a person or entire ethnic group are existentially inside place or existentially outside of place (Relph 1976; Seamon 1996; Seamon and Sowers 2008).

Kinship affects change through relations of custody and constraint, obligation and reciprocity. The same can be said for the kinship between people and place, when place is intimately linked to ethnic identity or becomes the context through which harm is done to another ethnic group. It is not uncommon for people to assert "I know the influence of place", acknowledging its constraint over human life, meanwhile passionately articulating its centrality in supporting the constitution of an ethnic identity. So too is place held responsible for healing and nurturing its human counterparts. A sense of kinship also informs the responsibilities people have for place and the dutiful behaviours that may be enacted in and for place. The obligation to care and desist in harmful practices that may bring about ill-effects in place mirrors anthropological understandings of the role of kinship, which serves to not simply organise human relations into familial forms, but create threads of connection between social beings and other presences. Kinship between people and place is constituted by regulative conceptions of social, political and economic life (Viveiros de Castro 1996: 183). Certain actions are deemed 'of and in place' or 'out of place' and are accompanied by an ongoing alertness to otherness-in-relation (Edminster 2011: 140).

Distinguishing place as agent and kin is not an exercise in anthropomorphising the place world. Instead it encourages awareness of the distinct ontologies through which place articulates its order and disorder, health or harm. As grappled with through animal studies, and the projection of human meaning onto non-human animal life, deeming place agency necessarily comparable to that encountered in human life might suffer a similar fate of misrepresentation. Thus, the methodology adopted here operates alongside the knowledge that if place were to speak, I and others may or may not understand what it has to say. This does not mean, however, that we are prevented from listening (see Rose 2013b; Hokari 2011) or that all epistemologies necessarily struggle with this matter of 'translation'. For there do exist countless indicators of agency contained in elements of the world that manifest in place, be they spatial features capable of contracting and expanding or climatic, geological, biological and ecological features that drum and move in sequence and for purpose or out of sequence and into chaos. These convey an ontological order that through repetition can be read for consistency and variation.

Evidence of place agency is not rare. It is widely documented in Indigenous testimonies, and ethnographic accounts of people and place (see Cajete 2000; Battiste and Youngblood Henderson 2000; Deloria 1994; Kahn 2011; Rose 1996; West 2005, 2006; Yunupingu 1997). These accounts reveal that place can be the source of spirit children, entities that are harboured by place until such time as a person is ready to be born. So too, mountains are known to move, while place may rush closer to the weary traveller who cries for their destination to arrive more quickly. Place may be described as quiet or wild, conveying something of its personality at a moment in time, relative to its biography and processes of human abandonment or inability to instate and accept relationships with a human counterpart of the right or wrong kind (Rose 2001: 117). So too, ethnographically it has been found that place can steal a person, hide them, cause their death or their slow spiritual and physical decline. Often the reason for this is that place contains ancestors, as both spiritual beings – the ghosts of deceased kin – and elemental presences. In sum, place is a co-presence made up entirely of agents capable of reading human behaviour and responding. If place is imagined in such ways, it becomes very difficult to envision space as a blank canvas given colour and expression by human consciousness. So too it is impossible to conceive of emptiness and solitude. One begins to suspect that everything is watching and potentially stimulated by our human presence and in turn we are physically and emotionally triggered to act and behave in particular ways. Place has far more agency than perhaps anthropology and human geography have yet been willing to put their disciplinary name to. Yet opening up the field for a discussion of place agency is made possible through my experiences as an anthropologist and ethnographer, and in particular my exposure to Indigenous knowledges pertaining to place.

Operationalising an Indigenous epistemology of place is done through a methodology of decolonising principles, which safeguards against the 'othering' of Indigenous knowledge or the speaking on behalf of Indigenous peoples. It is not dissimilar to Johnson's (2012: 829) work on a critical pedagogy of place, that "seeks to decolonize and reinhabit the storied landscape through 'reading' the ways in which Indigenous peoples' places [and place more broadly] and environments have been injured and exploited".

By combining the approaches of decolonising principles with Indigenous knowledges of place, nested ecology, kincentricity and emotional geography, agency and sentiency in place are acknowledged. Although focused on a vast number of cultural instances in which people and place have been harmed, as both Indigenous and non-Indigenous populations, this methodology employs Indigenous epistemologies as a guiding intellectual framework for place agency and sentiency. In particular it adopts a form of 'perspectivism', which is characteristic of Indigenous knowledge systems worldwide (Viveiros de Castro 1998). Viveiros de Castro (1998) details the perspectival quality of Indigenous thought, through specific reference to Amerindian cosmology, as follows: "the world is inhabited by different sorts of *subjects* or persons, human and non-human, which apprehend reality from distinct points of view". This approach is one of radical relationism whereby kinship, and other social phenomena reveal configurations of distinctions between humans and non-humans, which are irreducible to Western distinctions between nature and culture (Viveiros de Castro 1998).

More often applied to a discussion of human and non-human animal inter-actions, perspectivism and distinctivism are timely additions to a discussion of place-meaning. The actions of ancestral beings are, in many Indigenous cultures, responsible for the creation of place (Redmond 2001).[2] Places, once formed, are not static (Redmond 2001). They continue to have subjectivity and express intentionality in their constitutive parts, echoing Massey's (2005) view of space (as inclusive of place) as an ongoing production. On this topic, Redmond (2001: 121) writes about "features of the landscape that travel, shake, tremble and split". Far from being unexpected or atypical events, sto-ries of moving places fill the contemporary social universe for Indigenous nations (Redmond 2001: 122). Whether through being the embodiment of ancestors (or spirits) or the bounded and recognisable entity that acts as a pivot or anchorage point for a cultural group, there is a distinctiveness to place that may influence intersubjective relations. Distinctiveness is traced to the character of place and the internal form or logic that makes place what it is. Viveiros de Castro writes that:

> [I]ndigenous theory according to which the way humans perceive animals and other subjectivities that inhabit the world – gods, spirits, the dead, inhabitants of other cosmic levels, meteorological phenomena, plants, occasionally even objects and artefacts – differs profoundly from the way in which these beings see humans and see themselves.
>
> (Viveiros de Castro 1998: 470)

This theory postulates on the internal form of animals, and I, by extension, include place in this discussion. While not all animals (or places) are subject to this theory, those that perform a key symbolic and practical role within the group are regarded as having an internal form, which is constituted by a 'soul' or 'spirit' (Viveiros de Castro 1998: 471). This internal form is governed by intentionality and subjectivity in a manner that human consciousness also distinguishes itself (Viveiros de Castro 1998: 471). Thus, according to the perspectivism of Indigenous epistemologies, the elementary categories of social life organise relations between people and place, thus defining a social continuity between the two (Viveiros de Castro 1998: 473) This view differs from Western cosmologies, which suppose "an ontological duality between nature, the domain of necessity, and culture, the domain of spontaneity, areas separated by a metonymic discontinuity" (Viveiros de Castro 1998: 473). It becomes such that the space between people and place is itself social (deeply relational). Kinship provides the bridge between these distinct presences, both of which act with intentionality and subjectivity but that might come together through intersubjectivity. "Being and place are conceptually linked. This is an Indigenous principle, and, therefore, is maintained as such within Indigenous cultural philosophy and expressed in the most common or ordinary way" (Ortiz 2007: 135).

While Western paradigms of place-meaning might institute a split between the domains of people and place (or space), revealing distance between the social and 'natural' worlds, the approach adopted here shrinks this distance by emphasising the social continuity between the two and the agency of both. As Johnson (2012: 830) remarks: "Eurocentric thinking, broadly speaking, is based in a dualistic and reductionistic structure which divides the world into distinguishable segments with (supposedly) essential differences." By questioning these separations, and highlighting their intellectual origins as the 'Imperial binary' (Ashcroft 2001, cited in Johnson 2012: 830), social continuity between people and place can be reinstated through kinship and relationality, which, despite the distinctiveness of people and places, creates the possibility of both being active in the production of affect and responsive enough to feel the effect of one another. In taking this approach I am inspired, in part, by the work of Hokari (2011), who campaigns for the telling and writing of 'dangerous histories'. Seeking ways to explain what he had learnt in his practice as a historian learning Indigenous Australian histories, he encourages the reader to accept the danger of multiple ways of knowing and to sit in the gap that has been generated by the West in its failure to understand Indigenous epistemologies and ontologies.

Hokari (2011) confronted 'dangerous histories' through Indigenous ethnography as those that defy categories of good or bad (right or wrong in terms of disciplinary practice and legitimacy) and compels us to not simply add cross-cultural perspectives to our engagements with human life but to make our disciplines cross-cultural. Rather than incorporate plurality into a singular conceptual framework, he encouraged plurality in the very practice of learning to know something and understand something. This means changing the very way we see, look at and hear accounts of social life, and in this instance, place. In the vast number of place studies available to anthropology, and also cultural geography (and for Hokari in history) understandings become "logical and reasonable *only* if we stand and rely on Western/

universal-oriented historical [also anthropological and geographical] consciousness" (Hokari 2011: 253, emphasis in original). Deferring to Spivak (1988: 295), Hokari (2011: 254) confirms that an answer lies in the "systematic unlearning" of postcolonial intellectual privilege.

Writing to the presence and experiences of non-human agents, in this case place, tests me in this capacity to systematically unlearn and fully accept, not as alternative but as central, an Indigenous epistemology that configures place agency and sentiency. Place is now understood as alive, sentient, acting and responsive. How well I can achieve this is explored throughout this book via an examination of cultural wounding in place, using a methodology of Indigenous epistemologies of place, kincentricity, nested ecology and emotional geography. Perhaps there is a gap in what I can conceptualise, for my non-Indigenous identity is so deeply embedded in educational histories of the West,[3] but as Hokari (2011: 256; also Chakrabarty 2000: 108) encourages, staying with the gap might be the way to ensure communication between deeply plural spaces. Attempting this means working through a decolonising methodology and acknowledging ontological diversity, and taking seriously the philosophies of Indigenous peoples (Johnson 2012: 833). As a refreshing reminder of the work already done, Johnson (2012: 833) recalls:

> just as physicists ... have already done, social scientists and humanists are beginning to find bridges between modernism's binary reductionism and the holistic approaches common to Indigenous epistemologies. Anthropologists, environmental historians and eco-lit authors..., are attempting to bridge the divide between culture and nature, tearing down the city walls and its artificial liminality.
>
> (Johnson 2012: 833)

Harvey's (2006: 15) advice rings loudly in my ears: "More hopefully, when the Cartesian gulf begins to look like a hole someone dug, we begin to see that it is possible to climb out of it, fill it in, and walk away to do something more useful."

Decolonising methodologies of this kind have been inspired by the critiques of those who call out the unequal power that has long defined research and allowed for the alienating of Indigenous persons and their knowledge from intellectual inquiry and knowledge production (Smith 2012; Zavala 2013). Taking to task the intersecting worlds of Indigenous peoples and that of research, Smith (2012) and others (Battiste 2000; Rigney 1999; Zavala 2013) have reflected closely on the "absences, silences and invisibility" of certain people and ideas in the research process. Smith's work, first published in 1999 and then again in 2012, highlights the extent to which Indigenous knowledge is expert knowledge and has emerged as a vital intellectual space amid conditions that have largely conspired to silence it through pervasive and powerful canons of imperial and colonial thought. Indigenous knowledges and theory are not peripheral, never have been (despite being cast as such by coloniality and modernity), and feature as a vital part of this work because they make available a holistic and socially sophisticated framework for appreciating the relationship between people and place (Battiste and Youngblood Henderson 2000; Barnhardt and Kawagley 2005; McGregor 2004; Nakata and Langton 2009; Rose 2000; Smith 2012). In an ongoing effort to decolonise my own practice as an anthropologist and cultural geographer, I do not wish to reiterate the canons of 'Western' theoretical and philosophical approaches to place and place relations. I am seeking something more and feel a conversation of cultural wounding and ethnic violence demands this.

Indigenous epistemologies and place-meaning

> Everything is considered to be 'alive' or animated and imbued with 'spirit' or energy. A stone has its own form of animation and unique energy. Everything is related, that is, connected in dynamic, interactive, and mutually reciprocal relationships.
>
> (Cajete 2000: 75, 79)

> All things, events, and forms of energy unfold and infold themselves in a contextual field of the micro and macro universe.
>
> (Cajete 2004: 55)

Mindful and extremely cautious of essentialising Indigenous perspectives when referring to broad themes such as Indigenous epistemologies, ontologies and axiologies, the approach to place that is adopted throughout this work draws upon documented accounts of Indigenous place-meaning by Indigenous authors across time and space. There remains, however, the need for caution in the exercise of defining 'Indigenous knowledge'. To deny the plurality of Indigenous knowledges is not the aim of this work. Any effort to singularly define such knowledge would be, as Battiste and Youngblood Henderson (2000: 35) write, "loaded with Eurocentric arrogance". There is no blanketing concept or application of knowledge across Indigenous nations and communities. "It is a diverse knowledge that is spread throughout different peoples in many layers" (Battiste and Youngblood Henderson 2000: 35).

The discreteness of Indigenous knowledge is traced to the specifics of its emplacement. Each Indigenous nation or clan group emplaces their knowledge within specific contexts of meaning, which ensure that they can relate to their country, homelands and associated places of great importance. Thus, place knowledge reflects the distinct epistemology, ontology and axiology in which it is suspended. Rose (1996: 7) reminds us of Indigenous Australian relationships to place, and the extent to which knowledge is specifically catered to the 'country' and homelands in which it is found or to which it is linked.[4] Thus, place knowledge is nuanced, reflecting the character and identity of place, in as much as it does the character and identity of its people and knowledge holders. She writes

> People talk about country in the same way they talk about a person: they speak to country, sing to country, visit country, worry about country, feel sorry for country, and long for country. People say that country knows, hears, smells, takes notice, takes care, is sorry or happy. Country is not a generalised or undifferentiated type of place, such as one might indicate with terms like 'spending the day in the country' or 'going up the country'. Rather, country is a living entity with a yesterday, today and tomorrow, with a consciousness, and a will toward life.
>
> (Rose 1996: 7)

For insight into Indigenous knowledges to flourish, "scholars need to see Indigenous knowledge as a new *sui generis* (self-generating) path, as a new opportunity to develop greater awareness and to discover deeper truths about ecologies and their forces" (Battiste and Youngblood Henderson 2000: 39). My aim throughout this book is to fully embrace this principle, and not diminish Indigenous perceptions of place by definitional wrangling. I aim to, as best as I am capable, engage with key thoughts and intellectually agreed-to articulations of place knowledge, authored and shared by Indigenous people themselves. While often representative of localised settings and local knowledge, some themes emerge across the literature, and also in ethnographies and collaboratively constructed anthropological narratives of Indigenous place-meaning that assert an overarching theoretical paradigm. Cajete (1986:17–18; 2000) indicates that strands of connectedness do exist among Indigenous thought; a view balanced by Battiste and Youngblood Henderson's (2000: 41) reflection that "[g]iven the existing ecological diversity, a corresponding diversity of Indigenous languages, knowledge, and heritages exists". Utilising a set of ideas and philosophies generated by Indigenous knowledges across a broad range of contexts promotes the view that human relationships with place are always about active engagement and kinship. Some of the key principles include: knowledge of and belief in unseen powers; knowledge that all things contained in the world and ecosystem are dependent on one another; knowledge that personal relationships reinforce the bond between people, place and all other elements; and knowledge that order and disorder are relationally constituted between human and other presences through actions (Battiste and Youngblood Henderson 2000: 42–43).

This means that place itself is a dynamic and relational structure in which we are already embedded, rather than a static object over which 'ownership' or narrative control can be asserted (Malpas 2009). "[W]hile our connection to place operates through our engagement with and movement through place … we are ourselves inevitably shaped by those places, and so stand under their sway" (Malpas 2009). As with kin, "we are, one might say, 'owned' by place in a way that is quite different from any ownership we might attempt to exercise over particular places" (Malpas 2009). Shaw et al. (2006: 270) write that "this sense that humans have a kinship relationship with their environment is frequently expressed by indigenous speakers". They cite Sioux author Lame Deer, who explains "how the ceremonial use of the term 'all my relatives' linked the speaker not only to the others in the ceremony, but also to everyone and everything else". "Given this web of relationships and responsibilities, indigenous geographies – in the broader sense – have long been characterized by an overarching emphasis on place – the specific, the meaningful and the interrelated" (Lame Deer and Erdoes 1972, cited in Shaw et al. 2006: 270). In the seminal work *Wisdom Sits in Place* (Basso 1996: 70), Apache man Dudley, reflects: "Wisdom sits in places. It's like water that never dries up. You need to drink water to stay alive, don't you? Well, you also need to drink from places. You must remember everything about them." In describing a 'Native Science' approach to place, Tewa intellectual, Cajete (2000: 41) refers to an intellectual commitment that requires "mutual reciprocity, [and] which presupposes a responsibility to care for, sustain, and respect the rights of other living things, plants, animals, and the place in which one lives". The universe thus becomes a "living breathing entity", "considered to be 'alive', animate and imbued with 'spirit' or energy" (Cajete 2000: 41, 75).

Interdependence and kinship are highlighted in many Indigenous philosophies of place and the universe more broadly. As Battiste and Youngblood Henderson (2000: 9) remark in the opening pages of their work *Protecting Indigenous Knowledge and Heritage*, "from the beginning, the forces of the ecologies in which we live have taught Indigenous peoples a proper kinship order and have taught us how to have nourishing relationships with our ecosystems ... These ecologies do not surround Indigenous peoples; we are an integral part of them and we inherently belong to them" (Battiste and Youngblood Henderson 2000: 9). For Cajete (2000: 52), humans are related to plants, animals, stones, water, clouds and everything else surrounding them. This is reflective of one of the oldest ecological principles practiced by Indigenous people all over the world, past and present. If you depend upon a place for your life and livelihood, you have to take care of that place or suffer the consequences, a lesson learned and relearned by many generations over time (Cajete 2000). Another part of this relationship is the emphasis on participation, meaning that "[w]e, as human beings, cannot help but participate with the world. We are all active entities" (Waters 2004: 50, cited in Diggle 2005: 7–8) who constantly interpenetrate with the world around us at various levels: the air we breathe, the carbon dioxide we contribute to the food we transform, and the chemical energy we transmute at every moment of our lives from birth to death all contribute to a constant mutual exchange between us and the world. People are biologically and energetically intertwined with place as much as they are socially and politically responsible for or invested in it. Conveying a form of animism, coupled with perspectivism and distinctivism, place is alive, and imbued with its own character, spirit, mood and energy. The earth is perceived as a "living soul" (Cajete 2000: 186), as well as a "living, breathing entity" (Cajete 2000: 41), that has the power to move, create and destroy. Thus, in dealing with place, we face a formidable co-presence.

In a number of "Indigenous ecosophical traditions, a benevolent other-than-human being is involved in transforming the environment" (Johnson 2007: 43). Johnson writes that

> numerous reoccurring ecosophical themes appear in the oral traditions of Algonkian peoples, among them cooperation and reciprocity between the people and other-than-human beings and transformation of the landscape by other-than-humans.
>
> Whereas Western science may contend that glaciation carved out the landscape … Native Science encodes the event in a memorable, culturally significant form. These stories need not be contradictory; rather, as an illustration of knowledge coexisting, Western science complements the memorable narrative of the Indigenous oral tradition.
>
> (Johnson 2007: 43)

In Haudenosaunee (Iroquois) epistemology the landscape is populated with human beings, in addition to "animals and plants, water, earth, and sky; fire, wind, rain, sun, and snow; thunder, lightning, shooting stars, and eclipses; stone giants, dwarfs, horned serpents, flying heads, often arrayed in families and societies, interacting with each other and with humans" (Dennis 1993: 21, cited in Johnson 2007: 42). For Wabanaki people of the Missisquoi region of New England: "We were created out of the wood of a tree that still thrives here" (Johnson 2007: 45, from Bruchac 2003: 13). Relatedness is an overriding theme across these narrations of place origins and meaning. This stands in contrast often to a non-Indigenous or Western tendency to overlook relatedness and kinship with place, preferring separation and anthropocentrism. Frequently cited as a distinct way of defining and knowing place, many Indigenous epistemologies establish human life as born of place.

If one is born of place, and can trace their ancestry, spirit origins and essence to place, then kinship is at the heart of the people and place relationship. Referring to Indigenous Australian views on the start of human life, Malpas (1999: 14) writes: "The mother does not contribute to the ontological substance of the child, but rather 'carries' a life whose essence belongs, and belongs alone, to a site [place]. The child's core identity is determined by his or her place or derivation." Yanyuwa elders in the southwest Gulf of Carpentaria, northern Australia, have described similar themes (Bradley with Yanyuwa Families 2016; Bradley 2010; Yanyuwa Families et al. 2003). For Yanyuwa, *ardirri*, the spirit children that inhabit the land and sea, are understood as placed in the earth by ancestral beings. Yanyuwa exclaim, for example, *rra-mangaji rra-kuridi kanda-yibarranthaninya ardirri nguthundu barra baji Wumanthala*, which translates to "the Groper ancestor she placed spirit children there to the north in that place Wumanthala" (Bradley et al.

1992: 120). This spirit child, once born and inhabiting the body of a living person, is held to reside deep within the bones of an individual, these being the least corruptible part of a person (Bradley et al. 1992: 120; Bradley with Yanyuwa Families 2016: 399, 408). This spirit child, a constitutive part of place, hence resides in the bones of a Yanyuwa person, taking the mutual constitution of people and place to a profound level.

Dwyer (1996) and Rose (2001: 100), reflect further, that in the context of Indigenous place relations, all is "cultural", there is "no 'nature'". People live within their landscapes, with "the impress of them upon it and of it upon them" (Dwyer 1996: 162, cited in Rose 2001: 100; see also Yunupingu 1997). Conveying the living qualities of place, Rose (2001) describes the action and agency of ancestral beings that created Indigenous Australian territories and homelands. Everything that exists in the world is alive, "animals, trees, rains, sun, moon, some rocks and hills, and people are all conscious. So too are other beings such as the Rainbow Snake, the Hairy People, and the Stumpy Men. All have a right to exist, all have their own places of belonging, all have their own Law and culture" (Rose 1996: 23). Place is held to be bursting with life (Rose 1996: 24), it is richly animate, cognisant, and visceral, and described as:

> walking, slithering, crawling, flying, chasing, hunting, weeping, dying, birthing. They were performing rituals, distributing the plants, making the landforms and water, establishing things in their own places, making the relationships between one place and another. They left parts of themselves, looked back and looked ahead, and still travelled, changing languages, changing songs, changing skin.
>
> (Rose 2000: 104)

Places of ancestral activity make up the "sacred geography of Australia" sourced by the prevailing presence of non-human agency and sentiency (Rose 2001: 104). This geography, however, is not set in stone (Redmond 2001: 120). It moves, contracts, expands and finds only momentary balance, "far removed from the static, timeless images of land" that non-Indigenous peoples are inclined to imagine (Redmond 2001: 121). Writing on place movement as a form of agency, Redmond (2001: 120) recalls that when an elder Ngarinyin person from the north west Kimberley region of northern Australia, exclaimed "we humans did not make the world with our own hands", one realises that the processes that bring place and the universe into being are "not reducible to either the practical consciousness or the physical capabilities of individual human beings". Operating in this context are logics of place intentionality and place's transformative capacity. For these moveable presences (both place and ancestral beings), their agency and sentience lingers and holds on, even transmogrifies relative to a range of effecting conditions that shape place character, mood and particularism. For as Rose (2001: 104) writes, "sentience and agency are located all through the system: in human and non-human persons, in trees, rocks, stones, and hills; wherever Dreamings [ancestral beings] are, there sentience is".

An Indigenous perspective brings to this discussion a distinct view of relationality and kinship in and with place. But is it beyond a non-Indigenous reasoning to imagine such relationships with place? Could this not be a realisation for all humans who are willing to expand their perceptual capacity? The challenge in realising the greater effects of place and, in turn, cultural wounding, lies in applying the concept of place kinship as Rose describes it:

> relatedness is the meat of life, situating people's bodily presence in shared projects that link human and non-human interests around intersecting and crosscutting contexts of tracks, countries, totems and sites. Every discrete category is linked to other discrete categories through kinship, and is crosscut by other discrete categories; thus the concept of exclusivity is both sustained (because categories are discrete), and demolished (because they are crosscut). This system links species, places, and regions and leaves no region, place, species or individual standing outside of creation, life processes and responsibilities.
>
> (Rose 2000: 105)

So too, Malpas calls non-Indigenous intellectualising of place to action:

> if we [non-Indigenous persons] are even to begin to admit some recognition, in our own experience, of the presence of something of the Aboriginal feeling for the intimacy of the connection to land and to locality, then we cannot avoid trying to understand that which might provide the ground for such notions. We cannot avoid trying to think through the diversity of issues that surround the idea of place, and our belonging to it, in some sort of unitary fashion.
>
> (Malpas 1999: 14)

There is an emotional resonance that comes from all human connections to land, sea and elements. This conveys something of the intimacy of being in and of place, an experience that shapes the human mind and body but that also has the capacity to flow back into place and affect its character and own health. The extent to which this resonance is understood and appreciated as a form of kinship between people and place, however, differs vastly across cultural terrains and epistemic spaces. Accepting such possibilities is made easier by a human geography that is accessed via a phenomenological approach to people and place. Through a language of co-substantiation, people and place are bound across all cultures through the meaningful structures of cultural life. That these are so well articulated by Indigenous epistemologies is evident and traceable to a pervasive sense of culture and relatedness. Separation is not welcome in this discourse of people and place. Realising this should inspire broader conceptualisation of emotions and affect in place across a range of cultural contexts and knowledge systems.

Knowing the gap

The definition of place, as applied here, needs to exhibit minimal 'spatial anxiety', and be capable of encompassing vast distinctions of place as home, as threat, object and instrument, a grid point on a map or a set of memories and recollections. By minimal spatial anxiety, I am referring to a definition of place that recognises both its tangible and intangible qualities. Overcoming spatial anxiety means operating somewhere between two poles, acknowledging place as part of a spectrum of possibilities, on the one hand capable of being traced to a spatial, physical and environmental point, but also existentially available to us through the mind's eye. Broad by nature, this work settles on place distinctions that are capable of holding the vast ways in which people conceive of place, but most importantly those that face towards place agency and relationality. These may translate into place distinguished as definitive, bounded and unbounded, a position in ordered space (ordered intellectually, ecologically or physically), an environment with its own character and a site of dwelling or harbouring human life. Beyond this, the deeper philosophical workings on place and its relationship to space are left to the likes of Malpas (1999, 2009), Lowenthal (1985), Casey (1997), Relph (1976, 1981) and Massey (2005). Having intensively explored space and place, they have collectively drawn on Western philosophical principles reflected in the works of Bachelard (1964), Deleuze and Guattari (1987), Heidegger (1971, 2008 [1962]) and Foucault (1986), championing a human(e) geography. While led heavily by Western theory, these treatments of place have not been disinclined to take on board Indigenous and Eastern perspectives to flesh out the possibilities of place and the nature of people and place relations. Anthropology and cultural geography have benefited solidly from their efforts in seeking to carve out a cross-cultural and intercultural space in which to better understand the scope and range of human relations with place (see Low and Lawrence-Zuniga 2003; Rodman 1992).

Offering a review of place-meaning, Malpas (1999: 22, see also Johnson 2012) provides an appealing set of themes, which encapsulate several domains for place existence. Out of these five themes he surmises place to be (1) a definite but open space, (2) a more generalised sense of place, extension, dimensionality or 'room', (3) location or position within some order (spatial or otherwise), (4) a particular locale or environment that has a character of its own; and (5) an abode or that within which something exists or within which it dwells (Malpas 1999: 22). Malpas (1999: 22) acknowledges also that "whilst this summary captures many, it does not capture all the shades of meaning that place can carry". In response, I offer a sixth dimension to place, namely a dynamic and relational structure in which human life is embedded, thus making it the embodied location of everyday struggles for meaning, political, economic and cultural life (Johnson 2012: 830).

Seamon (2012: 3) offers further insight regarding how place might be understood as agential and pursued as a vital part of human life. He (2012: 3) writes of place "as an integral structure of human life", one that can be understood in terms of three dimensions: first, the *geographical ensemble* – as the material environment, including both its natural and human-made dimensions; second, *people-in-place*, including individual and group actions, intentions, and meanings; and, third, spirit of place, or *genius loci* – the prevailing character or atmosphere of a place. Imagining the ways in which place contributes to human life, Seamon (2012: 3) does not grapple with the notion of place agency, instead presenting it as tacit. If place is able to contribute to human life, therein lies the fact that relationality is at the heart of place-meaning. This relationality between people and place is facilitated by the following; place interaction, place identity, place creation, place intensification, place realisation and place release. *Place interaction* "refers to the typical goings-on in a place", it belies "a day in the life of a place and involves the constellation of more or less regular actions, behaviors, situations, and events that unfold in the typical days, weeks, and seasons of a place" (Seamon 2014: 16). *Place identity* "phenomenologically relates to the process whereby people living in or otherwise associated with a place take up that place as a significant part of their world", thus recognising the specific qualities of place and choosing to instate and maintain kinship with it (Seamon 2014: 17).

Place creation and intensification "speak to how positive human effort and well-crafted making can improve places or, through inappropriate understandings and constructions, can activate place decline" (Seamon 2014: 18–19). In place creation, human beings are active in relation to place, whereas, place intensification invokes place as the active agent in relation to human beings. *Place realisation* refers to the "palpable presence of place", while *place release* conveys something of the serendipitous nature of place, whereby, "through unexpected experiences, situations, and surprises relating to place, people are released more deeply into themselves" (Seamon 2014: 19). Place release describes the unexpected qualities that come about in place, triggered by the spontaneity of people in place and place's response to people. Seamon (2014: 17) provides examples of place release, including "meeting an old friend accidently on the sidewalk; enjoying the extemporaneous performance of an itinerant street musician; or becoming friendly with, dating, and eventually marrying the checkout clerk who just happened to take your take-out lunch order each work day". "Partly because of place, 'life is good'" (Seamon 2014: 17). Seamon (2014) reflects further, however, on the ways in which place release can lead to experiences of displeasure or uncertainty. Both imply agency beyond human influence. It is place that has the capacity to uplift or unsettle human life. In the case of the latter, this comes about when place less often or no longer offers enjoyable surprises and unexpectedness; users feel less a zest for daily life to which the place formerly contributed. In a more crippling mode, release as undermining place can involve disruptive serendipitous situations whereby one is upset or hurt – for example, one happens by chance to be mugged in front of the apartment house where he or she lives.

(Seamon 2014: 17)

Seamon's (2012: 3) work testifies to the fact that place identity exists and is vital to understanding the character of place and its terms for relating. It interacts heavily with human life, and not merely because people perceive this to be the case or imagine it through their cultural frameworks and intellectual apparatus. Interaction occurs because place is an agent; it does as it might wish and responds in particular ways.

By drawing together this kind of approach to place, I avoid explicitly separating out place from space. This separation, long held to adequately distinguish the 'real' or natural world (space) from the conceptual and social one, (place) defines space as the ultimate context of life over which we command less influence, that ether in which we float. Massey (2005: 6) grapples with the implications of this, questioning how it is we are left with a notion of space as "the outside? The abstract? The meaningless?" and place as "meaningful, lived and everyday". By such criteria space transformed by human meaning is what becomes envisioned as place; "a reflection" – the deeply human realm over which we have authorial control through perceptual capacity, which leads to knowing, naming and claiming of place identity and rights. When left with these two options, either space as the elemental universe that surrounds us, or place as the projected human construct, deeply intimate yet limited in potential by what we as enculturated humans are capable of perceiving, I cannot fully articulate the relational experience of cultural wounding that I seek to explore in a book of this kind.

For the most part, space has been approached geographically as 'something we humans pass through', a container within which other things happen (Massey 2005; Anderson 2005). The tendency has also been to conflate space with the non-human and environing aspects of biological and geological life; somehow "originarily regionalised" (Massey 2005: 6). Space has been described as politically neutral, fixed and dead, while place has satisfied the criteria of being overtly political. Questioning the distinction between space as politically neutral and place as socially constructed, Massey (2005) reflects that outside of constitutive relations, space has no existence. An argument for political neutrality cannot sustain, in that space is the "very ground of the political because to think spatially is to engage with the existence of multiple processes of coexistence. That is, it opens up a type of relational politics based on the negotiation of relations, configurations" (Massey 2005: 147). "Space is the product of *interrelations*", thus accordingly; "we must recognize space as constituted through interactions, from the immensity of the global to the intimately tiny" (Massey, 2005: 9). Space is a sphere "of dynamic sim-ultaneity, constantly disconnected by new arrivals, constantly waiting to be determined (and therefore always undetermined) by the construction of new relations" (Massey 2005: 9; see also Anderson 2005: 230).

In a radical and deeply appealing sense, Massey imagines the constitutive relations of space to be all manner of heterogeneous bits and pieces (that are simultaneously natural, social, political, economic and cultural) (Anderson 2005: 231). This resonates with and begins to meld into a view of place, imag-ined and constituted through relations. Social relations are held to distinguish space, and place, in ways that are culturally comprehensible. More often writ-ten of as a 'sense of place', through perceptual capacity people make distinc-tions and decisions about what is in and of space (Feld and Basso 1996). This is done by ascribing boundaries, names, and by associating the body with space in a particular fashion. Thus place is often held to be the "sphere of the everyday, of real and valued practices" (Massey 2005: 5). That meanings are made according to culture as the self-conscious distinctions that hold ethnic and other cultural groups together is central to human geography and the anthropology of place.

When distinguishing space and place, there is also the tendency to grant space inherent agency. Space is held to consist of the geological, biological and temporal elements of the universe that can and do exist without human life. Place, however, conventionally defined by efforts of social construction through perception, is considered dependent on human life for its existence, thus is deemed lacking in its own inherent agency. What I am drawn to is a blurring of these distinctions, namely space as inherent and pre-existing unto another 'order' or 'logic', and place as human insertion or carving of meaning from space; a kind of 'taming' of space. Place depicted as the inti-mate other of space then becomes configured as inherent, autonomous of human creation, yet linked closely to people through kinship. As agent, place is capable of acting and responding, it has character of its own, yet as kin it is dependent upon human life and thus pulses to the rhythms and nuances of social life as well. This definition of place is dependent on a combination of the externality that is granted to space and the relationality that is accorded to place as human creation. It follows Massey's approach to space and place, which generates the query, how do we understand the durability of particular places? And how do certain constellations of relations repeat and endure in place? (Anderson 2005: 234). Such questions assist the exploration of radical alterity from within a system of relational thought. That is, how to engage with relations that remain unknowable to some, undecided or indeterminate? The latter I take to mean place agency and the possibility of place being capa-ble of feeling harmed through its relational intimacy with those who are cul-turally wounded.

In a discussion of cultural wounding, these two possibilities merge to become an entire 'situation' or realm of action, both in a physical sense and also as a structural condition. It is through this situation that harm is conceived of and into this realm that it is delivered. The situation in which cultural wounding occurs is key to understanding the motivation to wound and also the conditions for healing. Wounds cannot be delivered without a realm for human contact, idea exchange, subsequent action and meaning-making. But this is not context as merely a stage for action and/or backdrop for resulting trauma. Rather it is a meaningful component in the doing of harm and also the receipt of cultural wounding. As a meaningful component, I propose that human life cannot be separated out from its spatial field as constituted through actual physical and intellectual spaces. These two fields of presence (physical and intellectual space) do not go unaffected by human action any less, as they constitute vital elements and spheres within the ecology of life. An appreciation of this comes to a fuller realisation when separations are bridged.

Defining both space and place by their relational logic suggests that accepting a holistic sense of interacting presences relieves us of the need to distinguish one concept from the other. As Anderson surmises, trajectories in human geography mean that few now

> conceptualize space as a 'container' in which other entities or processes happen. Instead, any space or place, from the intimate space of a body to the space of the globe, are precarious achievements made up of relations between multiple entities. Spaces have to, in other words, be made and remade because relations are processual.
>
> (Anderson 2005: 230)

What Anderson (2005) and Massey (2005) offer is a relational approach to theorising space and place, which submits and promises the ethos and politics of contemporary human(e) geography. Malpas (1999: 1) implicitly questions the separation of space and place, stating that "the human relation to the land, and to the environing world in general, is clearly not a relation characterised by an influence running just one direction". The environing world through which humans move influences us, and in turn we influence it. As Malpas (1999: 1) states, the "environment determines our activities and our thoughts – we build here rather than there because of the greater suitability of the site; the presence of a river forces us to construct a bridge to carry the road across; we plant apples rather than mangoes because the climate is too cold".

Malpas (1999: 1) expands the discussion of relationality by introducing our own *affectivity* as well as our ability to *effect* as key considerations in the meaning of place. Human affectivity hints at one part of what is key to this book. Accepting the possibility that humans are affected or changed in some sense by place is relatively commonplace in the field of geography and certainly human geography, and reveals the possibility of agency in space and place. Changing human disposition, as emotional responses, and physical behaviour, space and place are presences to which we respond (see also Massey 2005). If people are not solely responsible for the creation and authoring of space and place, the likelihood that they are pre-existing or self-determined spheres must be entertained, thus rendering both as co-presences with which we interact, commune, respond and establish or reject kinship. Massey, who inspires us to think differently of space, states:

> What is needed, I think, is to uproot 'space' from the constellation of concepts in which it has so unquestioningly so often been embedded (stasis; closure; representation) and to settle it among another set of ideas (heterogeneity; relationality; coevalness … liveliness indeed) where it releases a more challenging political landscape.
>
> (Massey 2005: 13)

This encourages settling on the notion of space/place as an agent of effect, not only in harbouring the geographical, biological and climatic realities to which humans respond, but also in socially prescribed ways that may induce certain behaviours. Behaviours come to make 'sense' in place due to the demands or expectations of that very place. Prescribed behaviour, designed to appease place and/or diffuse the likelihood of place acting against human presences, has been shown to exist in some cultures. Ethnographically, in working with Indigenous families in northern Australia I have encountered and recorded attitudes and practices that convey a cultural comprehension of place as having the capacity to hide food resources, allow areas to become dangerously overgrown or riddled with pest species and other dangerous elements. In these moments, and others, extending far back to an ancestral creation period, place reminds the human and non-human visitor of its authority and influence in the relationship. In turn, people effect change on place's constitutive parts through agency of their own. The relational aspect of this is best aligned with the principle of kincentricity, which positions people and place as part of an extended ecological family.

Emotional geography and nested ecology

In her book *Root Shock*, Mindy Fullilove declares, "displacement is *the* problem the twenty-first century must solve" (Fullilove 2004: 5, emphasis added). A confounding event in human life, it derails those who experience it as the fundamental disordering and fracturing of life, self and place relations. Whether loss of home from bushfire, residential transience experienced by those on the brink of homelessness or the silence of an ethnically cleansed village, displacement erodes the cohesiveness of self and place. It undermines meanings that are essential to a sense of well-being and security, not only for the integrity of human life, but all life. Writing of the ecology of loss induced by bushfires in vulnerable parts of Australia, Cox and Holmes (2000: 69), remark that "[s]uffering is inevitably connected to the disruption of meaning". The loss of meaning that comes with the destruction of place and its elements is conveyed throughout a series of oral testimonies recorded after a major fire event. One woman explains her sense of "terrible heaviness, a sadness for the bush and the birds" (Cox and Holmes 2000: 70). She continues to map place through the interactive co-presences that have been lost: "We used to go up the back and take bread and green stuff for the birds and anything that was left living … I found lots of 'roos [kangaroos] crouched, like parchment just as they'd died and a flock of wild geese in a hollow where they'd sheltered" (Cox and Holmes 2000: 70). Cox and Holmes (2000: 70) recall that "grief for her own losses came later, but was never as profound as her grief for the losses in the environment and these aspects of deep grief for animals and for the natural environment hinted at a belief in a universe where humans are part of, rather than superior to, nature". The profundity of place relations, and place as kin, are not lost in this account.

For those who imagine displacement, yet live with the secure knowledge of assured place relations, the thought provokes anxiety of unknowable fear and nightmarish loss. Fullilove encourages empathic concern for place and its human counterparts, reminding us that

> Africans and [A]borigines, rural peasants and city dwellers have been shunted from one place to another, as progress has demanded, "Land here!" or "People there!" In cutting the roots of so many people, we have destroyed language, culture, dietary traditions, and social bonds. We have lined the oceans with bones, and filled the garbage dumps with bricks.
> What are we to do?
>
> (Fullilove 2004: 5)

Plunging the depths of an emotional geography, worlds of pleasure and pain are declared through an account of 'root shock', "the traumatic stress reaction to the destruction of all or part of one's emotional ecosystem" (Fullilove 2004: 11). This ecosystem is the place-based world in which our human being is held and shaped. What Fullilove describes are the terrifying responses people may have in moments of place trauma. The effects of place trauma (as loss of home) on our being are taken to be profound, shaking human life to its core and destabilising the relationships that ensure a reckoning of place and emplacement. Fullilove describes this as *root shock*. Root shock is what takes hold when the foundations of our lives are shaken and disordered in ways that are difficult, if not impossible to comprehend. Alluding to the rootedness of human life in place, physically and conceptually, when the roots are damaged or shaken, ill-health, hardship and struggle to survive can become the lived experience of everyday life. This root shock, she writes,

> has important parallels to the physiological shock experienced by a person who, as a result of injury, suddenly loses massive amounts of fluid. Such a blow threatens the whole body's ability to function. The nervous system attempts to compensate for the imbalance by cutting off circulation to the arms and legs. Suddenly the hands and feet will feel cold and

damp, the face pale, and the brow sweaty. This is an emergency state that can preserve the brain, the heart, and other essential organs for only a brief period of time. If the fluids are not restored the person will die. Shock is the fight for survival after a life-threatening blow to the body's internal balance.

(Fullilove 2004: 11)

The bodily metaphor holds to ground the idea that the environment too, and place, can receive violence and suffer shocking imbalance. Harm done to place by way of aggressions, deliberate destruction, carving, cutting, removal and 'cleansing' ricochet into the lives and through the bodies of the people who happen to reside there, trace ancestry to place or carry responsibilities for place health. Human sickness, suffering and depression are often traceable to aggressions against place (see Davis 2005a, 2005b; Hessing et al. 2005: 320; Windsor and McVey 2005). These same aggressions cause disorder in place, by way of a rearrangement of place components and loss of meaning and integrity (see Heider 2005). Rearranging place through its destruction or evacuation, rescripting or denial threaten internal place balance as they do to the human body. To deny place its own root shock is a further aggression, magnifying the experience of trauma. An overly humanistic approach to wounding, and the shock of violence renders the experience of place as background, yet when looking upon torn landscapes, annihilated townships or the exploded spaces of mining activity, wounding is evident in the physical disorder of place. The recognition of root shock for both people and place is one way to expand our emotional geographies. It echoes a principle of nesting, whereby,

Your [human] position in that small, central, nested domain is not a state-ment of your centrality – or that of humankind – to the rest of the world. It simply reflects the pragmatic reality that as a human being you can only perceive the world from your own subjective experiences, taking into account your biologically derived perceptual and cognitive capabilities, and you can only interact with those progressively more expansive eco-logical domains through the use of your body and the tools that your body is capable of fashioning.

(Wimberley 2009: ix)

While Wimberley (2009) might imagine the human perceptual field as our great limitation in recognising multifarious agents and co-presences in place, there is potential to expand our perceptual capacity in ways that are mirrored by Indigenous place knowledge. It is also through an emotional geography and rhetoric of nesting that a further bridging of the gap between an Indigenous philosophy of place and Western philosophies of place can be achieved. Nesting acknowledges a richly layered universe; "a view that may come to slowly and irreversibly replace the more mechanistic, static, vertical, and linear models that have been foundational to modern thought" (Haught 2009: vii). However labelled, these philosophies of separation are often rampant in contexts of cultural wounding and contact zones where place theft and annihilation is done with only a momentary concern for the well-being of place and its human kin. Often described as negligible loss, perhaps collateral damage or inevitable destructions ushered in by 'progress', place wounding and trauma are often and tragically deemed explainable and a 'natural' part of the conflict contract.

Kincentricity and nesting, along with emotional geography, significantly reduce the likelihood of sustaining such a view, and interrupt enduring apathy and claims to disentangle people from place and vice versa. Emotional geographies characterise the biographies of all people, all cultures, all ethnic groups and all nations. Just as "each person has a [natural] place which is considered to be the zero-point of his [or her] personal reference system"[5] (Buttimer 1976: 284) when called upon to reflexively imagine the experience of place as one plunged into emotional fields of terror, cruelty, hatred or indifference, the prospect of deprivation and dislocation sets many of us adrift. Gaining in presence and intellectual influence since the early 2000s, emotional geography pivots on the premise that "[w]e live in worlds of pain or of pleasure; emotional environs that we sense can expand or contract in response to our experience of events, though there is rarely a clear or consistent sense of simple 'cause' and 'affect'" (Davidson and Milligan 2004: 524). In some respects, this brings us closer to appreciating an Indigenous philosophy of place, which has as its foundations a belief in interdependence, unseen powers, bonds and the knowledge that order and disorder are relationally constituted between human and other presences through actions (Battiste and Youngblood Henderson 2000: 42–43). If emotions are triggered in place, emotional responses generated by place or the emotional dispositions of people themselves bring about certain practices or refrain in place, then it goes without saying that webs of connection are at play here.

Emotion, as the affective state of consciousness that is experienced in engagements with the world may include a suite of dispositions, from love to anger, fear, bewilderment and loneliness. In the moment of cultural wounding, an emotional geography may take on sinister qualities akin to disgust, ridicule and vengeance, or rage, grief and isolation. Emotions are responsive and take on particular meanings in place. Emotional geographies make possible an emotio-spatial hermeneutic, in which "emotions are understandable, 'sensible' ... in the context of particular places" (Davidson and Milligan 2004: 524). This is because of the dialogue of social life that makes people and place sensible or illogical to one another. Emotional geographies pivot on interrelations, as do nested ecologies (Wimberley 2009: 6). In turn, nested ecology – or a nestled sense of place, people and life – might encourage sufficient sensitivity to all the dimensions that a truly integral vision of life requires (Haught 2009: vii).

A nested ecological approach ensures that the relational nature of people and place is emphasised, not simply as a constructed association, but as one that is a pervasive and an inherent part of everyday life. Accordingly, there is no sense of pre-existing space into which humans write their presence as place. Instead, place is recognised as alive, a life that exists in relation to human existence. It is an agent capable of response and action. For Wimberley (2009: 5), a nested ecology pivots on interrelations between realms, inclusive of the personal ecology, social ecology, environmental ecology and cosmic ecology. A fifth nest is that of spiritual ecology, what Wimberley (2009: 5–6) describes as the realm of religion, spiritualism, divinity, desire to know, faith, doubt and belief. It is held that "changes in one portion of the system can be felt throughout other systems" (Wimberley 2009: 7). Nested ecological domains are mutually reinforcing and supportive of one another. To that end, the nesting being referred to conveys an "interlocking set of systems that begin at the level of the self and progressively extend to encompass families, groups, communities, ecosystems, the biosphere, and beyond into the unfathomable reaches of the cosmos" (Wimberley 2009: 4).

Wimberley (2009) assumes a hierarchical quality to this nesting, yet for all intents and purposes in this book, hierarchy is abandoned in that it serves no great purpose to rank these realms differently. Instead the greatest utility of a nesting principle is found in its relational propensity. Nesting returns us to kinship through its emphasis on relations and ethical awareness, in that an ethical incentive "can flourish most naturally and spontaneously when persons understand that their actions and attitudes have bearing not only on the physical, social, biological and cosmic spheres of being but even on the ultimate environment in which they live and move and have their being" (Wimberley 2009: 11). An ongoing alertness to otherness-in-relation can be traced through a nested logic, in turn feeding into a discussion of the greater effects of cultural wounding and violence. For Wimberley (2009: 11), "[a] nested ecology is devoted not simply to the study of 'natural' life, 'species' life, 'human' life, or the life of 'ecosystems', it entails considering all life and all that is required to maintain and sustain it". By seeking to understand the greater effects of cultural wounding, this work must go beyond the impact on human life, and wade into the greater field of co-presences, nests and kin, that too might be implicated in traumatic events and enduring episodes of hardship and struggle.

Chapter overview

Agency and sentiency in place have always been distinguishing parts of Indigenous epistemologies, ontologies and axiologies (Battiste and Youngblood Henderson 2000; McGregor 2004; Johnson 2007; Rose 1996, 2001). So too is kinship between people and place a governing part of Indigenous social and political life. Through clan-based affiliations and associations, place is a deeply socialised presence, which calls upon its people to interact according to laws and social codes. Place, as consisting of a limitless number of possibilities, from bodies of water to home to ancestral tracts of land or entire regions, possesses a faculty through which the external world (or surrounding world) is apprehended. It is thus capable of witnessing human action, and is frequently held to have the capacity to respond to this. By extension then, is it possible that place also feels the consequences of human action, in particular action that is designed to harm and has violent intent? When place's human counterparts are harmed and culturally wounded, I propose that it feels the drastic impact of this.

This chapter has introduced the reader to the principle of relatedness, outlining the inspiring premise of an Indigenous epistemology of place. Adopted with respect and in alignment with a decolonising methodology, this epistemology opens up the necessary space in which to reconfigure the people and place relationship as one grounded in kinship and relational logic. I recognise too that from within Western place studies, and human and cultural geography there are key concepts that might assist this study, namely emotional geography and nested ecology. I have sought to weave together this methodology as a tapestry not only designed to access multiplicities in place-meaning, but also to locate place as a co-presence in everyday life, and certainly in episodes of violence and cultural wounding.

The human experience of wounding and healing is written loudly into our social consciousness, whether grasped for its magnitude or dismissed by ongoing violence, structural amnesia and prescriptive forgetting (Connerton 2008). Despite this consciousness, surprisingly little has been documented on how place itself experiences violence and trauma. By nesting a view of people and place, one is seen to hold the other and vice versa. In which case, separation is futile, and the making of registers of greater or lesser suffering, or greater or lesser need to heal, are of no use here. This work accepts the view that amid violence and cultural wounding there is much scope for suffering and much need for healing. Seeking ways to explain this in finer detail, reflect on its contemporary impact and identify new principles for healing, recuperation and reparation are important steps in reconnecting presences in life that are too often separated. Ideologies of separation stymie healing as they often address only one aspect of what has been wounded. This chapter is an attempt to find a bridge that may link people and place in a more profound way. The rest is taken up throughout the remaining chapters where the specifics of place harm are outlined, the methods and motivations of place harm are diagnosed and pathways towards healing place are progressed.

In the following chapter, I provoke the experience and practice of witnessing place harm as part of a wider project of cultural wounding. This is done by introducing the reader to Yanyuwa country, as Indigenous homelands in northern Australia. Through ethnographic encounters with Yanyuwa and their homelands, I have witnessed the beauty and decline of ancestral places. The story of violence in place is rendered here through an exploration of these ethnographic accounts and other documented accounts of place harm and decline. By sitting in place, as wisdom does (Basso 1996), a deeper understanding of place harm and the wider ecology of wounding can be achieved. The task for Chapter 2 is to bridge the gap in ways of knowing place by engaging place itself as a legitimate primary source of wounding.

Notes

1 Culture, as it is used here, includes the self-conscious distinctions that are made by people who wish to share in a process of meaning-making about the self and the other (Kearney 2014: 3; Ewing 1990: 251). Culture of this kind designates spaces in which groups might enact these distinctions while more or less continuously changing. Place – both as physical presence and ideological map – is closely linked to the creation of an ethnic identity, as it is key to the substantiating of that identity through recursive links of being, knowing and behaving. Place becomes an essential part of validating and anchoring these identities into the world. Yet place is also a presence in human life that is ever-emerging in its spatial, biological and ideological character. There is no stasis in the realm of human cultural life, nor the place world we occupy. It is widely accepted that places change in their form, meaning and value. The catalyst for transformation determines how well these changes might be accommodated by place and place elements. Places change, both by nature and also through no fault of their own, hence they are as vulnerable to attack as ethnic groups, capable of being annihilated in physical and ideological terms.

2 If not the action of ancestral beings, there are frequently originary narratives of how people come to be emplaced in a particular way, thus highlighting the existence of place prior to the arrival and embedding of human life.

3 Yet this is offset by 16 years of ethnography in northern Australia and the teachings of Indigenous families, along with nine years of ethnography with African descendants in Brazil.

4 'Country' is a term used by many Indigenous Australian language groups to refer to their homelands as made up of land, sea, bodies of water, kin and resources. It is a holistic term. Rose (2014: 435) describes it further as an "Aboriginal English term", an "area associated with a human social group, and with all the plants, animals, landforms, waters, songlines, and sacred sites within its domain. It is homeland in the mode of kinship: the enduring bonds of solidarity that mark relationships between human and animal kin also mark the relationships between creatures and their country."

5 That said I am aware of liminality in people and place relations. Liminality in this case is that quality of ambiguity and/or disorientation that occurs in the middle stage of a sequence, or that becomes an experience in its entirety. When the person no longer holds an earlier state, disposition, affiliation, context or schema for meaning, but has not yet begun to transition into another. It is the standing at a threshold between multiple elements of the self in place, what once made sense, what once was available and what no longer can be made real or present in the life. Not necessarily a space of emptiness and partiality, but that moment when reconfiguring begins or disorder might be the prevailing experience.

References

Abujidi, N. 2014. *Urbicide in Palestine: Spaces of Oppression and Resilience*. Abingdon, Oxon: Routledge.

Alexander, J. 2004a. Toward a Theory of Cultural Trauma, in *Cultural Trauma and Collective Identity*. Edited by J. Alexander, R. Eyerman, B. Giesen, N. Smelser and P. Sztompka, pp. 1–30. Berkeley, California: University of California Press.

Alexander, J. 2004b. On the Social Construction of Moral Universals: The "Holocaust" from War Crime to Trauma Drama, in *Cultural Trauma and Collective Identity*. Edited by J. Alexander, R. Eyerman, B. Giesen, N. Smelser and P. Sztompka, pp. 196–263. Berkeley, California: University of California Press.

Alexander, J. 2011. *Trauma: A Social Theory*. Cambridge, Polity Press.

Alexander, J., Eyerman, R., Giesen, B., Smelser, N. and Sztompka, P. (eds.). 2004. *Cultural Trauma and Collective Identity*. Berkeley, California: University of California Press.

Anderson, B. 2005. Doreen Massey 'For Space', in *Key Texts in Human Geography*. Edited by P. Hubbard, R. Kitchin and G. Valentine, pp. 227–235. London: Sage.

Ashcroft, B. 2001. *Post-Colonial Transformation*. London, and New York: Routledge.

Bachelard, G. 1964. *The Poetics of Space*. Translated by M. Jolas. Boston, Massachusetts: Beacon Press.

Barnhardt, R. and Kawagley, A.O., 2005. Indigenous Knowledge Systems and Alaska Native Ways of Knowing. *Anthropology and Education Quarterly* Vol. 36(1): 8–23.

Basso, K. 1996. *Wisdom Sits in Places: Landscape and Language Among the Western Apache*. Arizona: University of New Mexico Press.

Battiste, M. 2000. Introduction: Unfolding the Lessons of Colonization, in *Reclaiming Indigenous Voice and Vision*. Edited by M. Battiste, pp. xvi–xxx. Vancouver, British Columbia: UBC Press.

Battiste, M. and Youngblood Henderson, J., 2000. *Protecting Indigenous Knowledge and Heritage: A Global Challenge*. Saskatoon, Canada: Purich Publishing Ltd.

Bracken, P. 2002. *Trauma: Culture, Meaning, Philosophy*. London: Whurr Publications.

Bradley, J. 2001, Landscapes of the Mind, Landscapes of the Spirit: Negotiating a Sentient Landscape, in *Working on Country: Indigenous Environmental Management of Australia's Lands and Coastal Regions*. Edited by R. Baker, J. Davies and E. Young, pp. 295–304. Melbourne: Oxford University Press.

Bradley, J. 2010. *Singing Saltwater Country: Journey to the Songlines of Carpentaria*. Crows Nest, New South Wales: Allen and Unwin.

Bradley, J. with Kirton, J. and the Yanyuwa Community. 1992. *Yanyuwa Wuka: Language from Yanyuwa Country*. Unpublished document, available at: https://espace.library.uq.edu.au/view/UQ:11306/yanyuwatotal.pdf. Accessed 26 November 2015.

Bradley, J. with Yanyuwa Families. 2016. *Wuka nya-nganunga li-Yanyuwa li-Anthawirriyarra: Language for Us, the Yanyuwa Saltwater People: A Yanyuwa Encyclopedic Dictionary*. Melbourne: Australian Scholarly Publishing.

Bruchac, M. 2003. Native Land Use and Settlements in the Northeastern Woodlands. Unpublished paper.

Buttimer, A. 1976. Grasping the Dynamism of Lifeworld. *Annals of the Association of American Geographers* Vol. 66: 277–292.

Cajete, G. 1986. *Science: A Native American Perspective. A Culturally Based Science Education Curriculum*. Unpublished doctoral thesis, International College, Los Angeles.

Cajete, G. 2000. *Native Science: Natural Laws of Interdependence*. Santa Fe: Clear Light Publishers.

Cajete, G. 2004. Philosophy of Native Science, in *American Indian Thought*. Edited by A. Waters, pp. 45–57. Malden, Massachusetts: Blackwell Publishing.

Casey, E. 1997. *The Fate of Place: A Philosophical History*. Berkeley, California: University of California Press.

Chakrabarty, D. 2000. *Provincializing Europe: Postcolonial Thought and Historical Difference*. Princeton, New Jersey: Princeton University Press.

Connerton, P. 2008. Seven Types of Forgetting. *Memory Studies* Vol. 1(1): 59–71.

Cox, H.M. and Holmes, C.A. 2000. Loss, Healing and the Power of Place. *Human Studies* Vol. 23: 63–78.

Davidson, J. and Milligan, C. 2004. Embodying Emotion, Sensing Space: Introducing Emotional Geographies. *Social and Cultural Geography* Vol. 5(4): 523–532.

Davis, J. 2005a. Representing Place: 'Deserted Isles' and the Reproduction of Bikini Atoll. *Annals of the Association of American Geographers* Vol. 95(3): 607–625.

Davis, J. 2005b. "Is It Really Safe? That's What We Want to Know": Science, Stories, and Dangerous Places. *The Professional Geographer* Vol. 57(2): 213–221.

Deleuze, G. and Guattari, F. 1987. *A Thousand Plateaus: Capitalism and Schizophrenia*. Translation and Foreword by B. Massumi. London and Minneapolis: University of Minnesota Press.

Deloria, V. 1994. *God Is Red: A Native View of Religion*. Golden, Colorado: Fulcrum Publishing.

Dennis, M. 1993. *Cultivating a Landscape of Peace: Iroquis–European Encounters in Seventeenth-Century America*. New York: Cornell University Press.

Diggle, S. 2005. Gregory Cajete, Traditional Santa Clara Philosopher. *APA Newsletters, Newsletter on American Indians in Philosophy* Vol. 4(2): 6–11.

Dwyer, P. 1996. The Invention of Nature, in *Redefining Nature: Ecology, Culture and Domestication*. Edited by R. Ellen and K. Fukui, pp. 157–186. Oxford: Berg.

Edminster, A. 2011. Interspecies Families, Freelance Dogs, and Personhood: Saved Lives and Being One at an Assistance Dog Agency, in *Making Animal Meaning*. Edited by L. Kalof and G. Montgomery, pp. 127–134. East Lansing, Michigan: Michigan State University Press.

Ewing, K. 1990. The Illusion of Wholeness: Culture, Self and the Experience of Inconsistency. *Ethos* Vol. 18(3): 251–278.

Eyerman, R., Alexander, J. and Butler Breese, E. (eds.). 2011. *Narrating Trauma: On the Impact of Collective Suffering*. Boulder, Colorado: Paradigm.

Falah, G. 1996. The 1948 Israeli–Palestinian War and its Aftermath: The Transformation and De-Signification of Palestine's Cultural Landscape. *Annals of the Association of American Geographers* Vol. 86(2): 256–285.

Fassin, D. and Rechtman, R. 2009. *The Empire of Trauma: An Inquiry into the Condition of Victimhood*. Princeton, New Jersey: Princeton University.

Feld, S. and Basso, K. (eds.). 1996. *Senses of Place*. Santa Fe, New Mexico: School of American Research Press.

Foucault, M. 1986. Of Other Spaces. *Diacritics* Vol. 16: 22–27.

Fullilove, M.T. 2004. *Root Shock: How Tearing Up City Neighborhoods Hurts America and What We Can Do About It*. New York: Ballantine Books.

Harvey, G. 2006. Animals, Animists, and Academics. *Zygon* Vol. 41(1): 9–20.

Haught, J. 2009. Foreword, in *Nested Ecology: The Place of Humans in the Ecological Hierarchy*. Edited by E. Wimberley, pp. vii–viii. Baltimore, Maryland: Johns Hopkins University Press.

Heidegger, M. 1971. Building, Dwelling, Thinking, in *Poetry, Language, Thought*, pp. 143–161. Translated by A. Hofstadter. New York: Harper Colophon Books.

Heidegger, M. 2008 [1962]. *Being and Time* (7th edition). New York: Harper and Row Publishers.

Heider, F. 2005. Violence and Ecology. *Peace and Conflict: Journal of Peace Psychology, Special Issue: Military Ethics and Peace Psychology: A Dialogue* Vol. 11(1): 9–15.

Hessing, M., Raglon, R. and Sandilands, C. 2005. *This Elusive Land: Women and the Canadian Environment.* Vancouver, British Columbia: University of British Columbia.

Hokari, M. 2011. *Gurindji Journey: A Japanese Historian in the Outback.* Sydney: University of New South Wales Press.

Johnson, D.M. 2007. Reflections on Historical and Contemporary Indigenous Approaches to Environmental Ethics in a Comparative Approach. *Wicazo SA Review* Vol. 22(2): 23–55.

Johnson, J.T. 2012. Place-Based Learning and Knowing: Critical Pedagogies Grounded in Indigeneity. *GeoJournal* Vol. 77: 829–836.

Jones, O. and Cloke, P.J. 2002. *Tree Cultures: The Place of Trees and Trees in their Place.* Oxford, New York: Berg.

Kahn, M. 2011. *Tahiti Beyond the Postcard: Power, Place, and Everyday Life.* Seattle and London: University of Washington Press.

Kearney, A. 2009a. *Before the Old People and Still Today: An Ethnoarchaeology of Yanyuwa Places and Narratives of Engagement.* North Melbourne: Australian Scholarly Publishing.

Kearney, A. 2009b. Homeland Emotion: An Emotional Geography of Heritage and Homeland. *International Journal of Heritage Studies* Vol. 15(2–3): 209–222.

Kearney, A. 2014. *Cultural Wounding, Healing and Emerging Ethnicities: What Happens when the Wounded Survive?* New York: Palgrave MacMillan.

Kelly, R. 1993. *Constructing Inequality: The Fabrication of a Hierarchy of Virtue Among the Etoro.* Ann Arbor: University of Michigan Press.

Lame Deer, J. and Erdoes, R. 1972. *Lame Deer: Seeker of Visions.* New York: Washington Square Press.

Low, S. and Lawrence-Zuniga, D. 2003. *The Anthropology of Space and Place: Locating Culture.* Malden, Massachusetts: Blackwell.

Lowenthal, J. 1985. *The Past is a Foreign Country.* Cambridge: Cambridge University Press.

Malpas, J. 1999. *Place and Experience: A Philosophical Topography.* Cambridge: Cambridge University Press.

Malpas, J. 2009. Place and Human Being. *Environmental and Architectural Phenomenology Newsletter* Vol. 20(3): 19–23.

Massey, D. 2005. *For Space.* London: Sage.

McGregor, D. 2004. Indigenous Knowledge, Environment and Our Future. *American Indian Quarterly* Vol. 28(3/4): 385–410.

Merlan, F. 1998. *Caging the Rainbow: Places, Politics, and Aborigines in a North Australian Town.* Honolulu Hawaii: University of Hawaii Press.

Nakata, M. and Langton, M. (eds.). 2009. *Australian Indigenous Knowledge and Libraries.* Sydney: University of Technology Sydney ePress.

Ortiz, S. 2007. Indigenous Language Consciousness: Being, Place and Sovereignty, in *Sovereign Bones: New Native American Writing.* Edited by E. Gansworth, pp. 135–148. New York: Nation Books.

Plumwood, V. 2002. *Environmental Culture: The Ecological Crises of Reason.* London: Routledge.

Porteous, J.D. and Smith, S.E. 2001. *Domicide: The Global Destruction of Home.* Quebec, Canada: McGill Queens University Press.

Redmond, A. 2001. Places that Move, in *Emplaced Myth: Space, Narrative, and Knowledge in Aboriginal Australia and Papua New Guinea.* Edited by A. Rumsey and J. Weiner, pp. 120–138. Honolulu: University of Hawaii Press.

Relph, E. 1976. *Place and Placelessness.* London: Pion.

Relph, E, 1981. *Rational Landscapes and Humanistic Geography.* New York: Barnes and Noble.

Relph, E. 1985. Geographical Experiences and Being-in-the-World, in *Dwelling, Place and Environment: Towards a Phenomenology of Person and World.* Edited by D. Seamon and R. Mugerauer, pp. 15–31. New York: Columbia University Press.

Rigney, L.I. 1999. Internationalization of an Indigenous Anti-Colonial Cultural Critique of Research Methodologies: A Guide to Indigenist Research Methodology and its Principles. *Wicazo Sa Review* Vol. 14(2): 109–121.

Rodman, M. 1992. Empowering Place: Mulitlocality and Multivocality. *American Anthropologist* Vol. 94(3): 640–656.

Rose, D.B. 1996. *Nourishing Terrains: Australian Aboriginal Views of Landscape and Wilderness.* Canberra: Commonwealth of Australia.

Rose, D.B. 2000. *Dingo Makes Us Human: Life and Land in Australian Aboriginal Culture.* Cambridge: Cambridge University Press.

Rose, D.B. 2001. Sacred Site, Ancestral Clearing, and Environmental Ethics, in *Emplaced Myth: Space, Narrative, and Knowledge in Aboriginal Australia and Papua New Guinea.* Edited by A. Rumsey and J. Weiner, pp. 99–119. Honolulu: University of Hawaii Press.

Rose, D.B. 2004. *Reports from a Wild Country: Ethics for Decolonisation.* Sydney: UNSW Press.

Rose, D.B. 2013a. Val Plumwood's Philosophical Animism: Attentive Interactions in the Sentient World. *Environmental Humanities* Vol. 3: 93–109.

Rose, D.B. 2013b. Death and Grief in a World of Kin, in *The Handbook of Contemporary Animism.* Edited by G. Harvey, pp. 137–147. Durham: Acumen.

Rose, D.B. 2014. Arts of Flow: Poetics of 'Fit' in Aboriginal Australia. *Dialectical Anthropology* Vol. 38(4): 431–445.

Sather-Wagstaff, J. 2011. *Heritage That Hurts: Tourists in the Memoryscapes of September 11.* Walnut Creek, California: Left Coast Press.

Seamon, D. 1996. A Singular Impact: Edward Relph's *Place and Placelessness. Environmental and Architectural Phenomenology Newsletter* Vol. 7(3): 5–8.

Seamon, D. 2012. Place, Place Identity, and Phenomenology: A Triadic Interpretation Based on J.G. Bennett's Systematics, in *The Role of Place Identity in the Perception, Understanding, and Design of Built Environments.* Edited by H. Casakin and F. Bernardo, pp. 3–21. Sharjah: Bentham Books.

Seamon, D. 2014. Place Attachment and Phenomenology: The Synergistic Dynamism of Place. Draft chapter online, in *Place Attachment: Advances in Theory, Methods and Research.* Edited by L. Manzo and P. Devine-Wright, pp. 11–22. New York: Routledge.

Seamon, D. and Sowers, J. 2008. Place and Placelessness, Edward Relph, in *Key Texts in Human Geography.* Edited by P. Hubbard, R. Kitchen, and G. Vallentine, pp. 43–51. London: Sage.

Shaw, W., Herman, R.D.K. and Dobbs, R.G. 2006. Encountering Indigeneity: Re-imagining and Decolonizing Geography. *Geografiska Annaler, Series B, Human Geography* Vol. 88(3): 267–276.

Smith, L.T. 2012. *Decolonizing Methodologies: Research and Indigenous Peoples*. New York: Palgrave.

Spivak, G.C. 1988. Can the Subaltern Speak?, in *Marxism and the Interpretation of Culture*. Edited by C. Nelson and L. Grossberg, pp. 271–313. Urbana: University of Illinois Press.

Trigg, D. 2009. The Place of Trauma: Memory, Hauntings and the Temporality of Ruins. *Memory Studies* Vol. 2(1): 87–101.

Tuan, Y.F. 1974a. Space and Place: Humanistic Perspective. *Progress in Geography* Vol. 6: 266–276.

Tuan, Y.F. 1974b. *Topophilia*. New York: Prentice-Hall.

Tuan, Y.F. 1977. *Space and Place: The Perspective of Experience*. Minneapolis, Minnesota: University of Minnesota Press.

Tumarkin, M. 2005. *Traumascapes: The Power and Fate of Places Transformed by Tragedy*. Carlton, Victoria: University of Melbourne Press.

Viveiros de Castro, E. 1996. Images of Nature and Society in Amazonian Ethnology. *Annual Review of Anthropology* Vol. 25: 179–200.

Viveiros de Castro, E. 1998. Cosmological Deixis and Amerindian Perspectivism. *The Journal of the Royal Anthropological Institute* Vol. 4(3): 469–488.

Waters, A. (ed.). 2004. *American Indian Thought*. Malden: Blackwell Publishing.

West, P. 2005. Translation, Value and Space: Theorizing the Ethnographic and Engaged Environmental Anthropology. *American Anthropologist* Vol. 107(4): 632–642.

West, P. 2006. *Conservation is Our Government Now: The Politics of Ecology in Papua New Guinea*. Durham, North Carolina: Duke University Press.

Wimberley, E. 2009. *Nested Ecology: The Place of Humans in the Ecological Hierarchy*. Baltimore: John Hopkins University Press.

Windsor, J. and McVey, J. 2005. Annihilation of Both Place and Sense of Place: The Experience of the Cheslatta T'En Canadian First Nation witin the Context of Large-Scale Environmental Projects. *The Geographical Journal* Vol. 171(2): 146–165.

Yanyuwa Families, Bradley, J. and Cameron, N. 2003. *Forget About Flinders: An Indigenous Atlas of the Southwest Gulf of Carpentaria*. Canberra: Australian Institute of Aboriginal Studies.

Yunupingu, G. (ed.) 1997. *Our Land Is Our Life: Land Rights: Past, Present and Future*. St Lucia, Queensland: University of Queensland Press.

Zavala, M. 2013. What Do We Mean By Decolonizing Research Strategies? Lessons from Decolonizing, Indigenous Research Projects in New Zealand and Latin America. *Decolonization: Indigeneity, Education and Society* Vol. 2(1): 55–71.

Part II

Witnessing place violence and the intent to harm

Part II of this book encourages the reader to witness place harm, to sit amid the pain, observe the destruction and more closely come to appreciate the experience of wounding for place and its relational world. By delving into the genealogy of a place's harm and possible death, we are compelled to recognise the actuality of place harm and the effects of violence on place and its people. We witness as readers, and through ethnographic accounts the chapter tells the story of place harm through the voices of those people for whom place is kin. While I cannot engage the testimony of place itself, although I would argue that this is accessible through displays of order and disorder, and other communicative events, it is hoped that by reflecting on 15 years of ethnography, and sticking close to the testimonies of Yanyuwa persons that have also been recorded since the 1980s, there is found an access point into place consciousness and being.

Yanyuwa are the Indigenous owners of land and sea territories in the southwest Gulf of Carpentaria, northern Australia, and are kin to a place world that is fully saturated with the presences of the ancestors as creation beings, non-human animals and the elemental parts of biological and geological existence. These elements are nested and come together to form the identity and character of a Yanyuwa world of meaning. The intention here is to present this discussion of place harm in a manner that encourages witnessing and listening to place in the midst of violence. Once violence in place is witnessed, then it is possible to undergo a process of diagnosing place harm, critically engaging with the methods and motives that underscore violent acts against people and places that matter.

2 An ethnography of place harm

This is a story of awful place harm. It began some time ago, and over the years has taken twists and turns, but never has the plot changed. The narrative of death, destruction and elemental decay has clung hard to this place. It is a big place, known as Bing Bong, but it is more complex than this name might convey. It is also a composite of places, scattered along the coastal margins of the southwest Gulf of Carpentaria, northern Australia (see Figure 2.0). In the local Indigenous language, Yanyuwa, these places are referred to as Makukula, Mawuli, Arrinyanda, Wimanda and Wurrulwiji. Each place is an important, living and breathing presence. Each has been formed by ancestral beings, which continue to reside in country, giving place its form and function. Due to the events that have occurred on this part of Yanyuwa country, many fear that these places are now dead or that they are slowly dying. Indeed many of the human kin associated with this area have died, first massacred by rogue settlers in the early colonial period, and in recent decades passing away from illness or accident, their demise believed to be linked to the decline of health in ancestral co-presences and as a result of the damage done to their ancestral bodies and esteems. Wilangarra and Yanyuwa, the Indigenous owners and kin for this place, didn't ask for any of this to occur, nor have they contributed to or accepted the decline of these profoundly important places.

Wilangarra, the original Indigenous owners for this part of the Gulf of Carpentaria, were, by the late 1800s, decimated entirely through murder and massacre, thus taken from their home territories in the most violent of ways. For Yanyuwa, who survived colonial violence, this part of the Gulf region was absorbed into their territorial reach. Several generations have now fought against the violence that has sought to and succeeded in wrecking place. This sustained resistance has been part of a difficult and exhausting fight against a formidable opponent, for whom the motivation was a colonial desire to rid this country of its rightful Indigenous owners and kin, and also to destroy it by contorting its shape, tearing at its features, digging up its innards and toxifying its lifelines – the latter done in the hope of making a financial profit. As a result, Bing Bong and all of its associated places have been left in a terrible state. In this chapter we are called upon to witness this violent story and consider how people and place as elemental co-presences might live through this. How they might begin to simultaneously heal from such experiences is of pressing concern, but requires that first we come to know the nature of the violence, the depth of the wounding and how it is held today in Yanyuwa epistemology, ontology and axiology.

Yanyuwa man Ronnie Miller describes the violence enacted against Bing Bong as causing deep wounds and a form of root shock that has left "country bleeding out". This is happening because country has "been cut so deep that it is bleeding out all the power of the old people" (pers. comm. with Bradley, 3 October 2015). 'Old people' is a nuanced expression referring to deceased kin whose spirits reside in place, returning to the clan country from which their spirit child originates. These old people, like ancestral beings, determine the order of place. The expression 'old people' is a distinction given to the most important of multifarious agents that exist in Yanyuwa country. They embody its Law and interact with their human kin through communicative expressions that are both subtle and overt, and capable of being read by Yanyuwa as part of a kincentric and nested ecology. Here, place is not space given a social veneer, it is the body and essence of ancestral beings. It is the substance of deceased kin and also an interacting co-presence that is keenly aware of disorder as it descends through acts of violence that intend to harm. Disordering acts are broad-ranging in their character, and for Bing Bong have come in the form of genocide, alienation and destruction, social disorder through renaming, toxicity and elemental erasure. The fear that these important parts of country are dying as a result of combined and long-term effects of cultural wounding and place harm is a deeply unsettling aspect of the Yanyuwa emotional geography.

Meeting Yanyuwa country

Yanyuwa are the Indigenous owners of lands and waters throughout the southwest Gulf of Carpentaria, northern Australia. While the whole of Yanyuwa country has been subject to the violence of colonialism, this chapter dwells on and in Bing Bong, as including Makukula, Mawuli, Arrinyanda, Wimanda and Wurrulwiji, and the events that have occurred here specifically. The violence here is symptomatic of Indigenous experiences around the world, yet they are localised in this discussion, thus allowing a more intimate reflection on the emotional geography of place and the greater effects of cultural wounding as part of a nested ecology. Bing Bong is on the western periphery of Yanyuwa country. It buffers alongside the homelands of another Indigenous language group the Mara whose own country continues to the west (see Figure 2.0).

The widespread killing of Wilangarra people was the first act in a series of deep woundings to have occurred here. Wilangarra are often absent from linguistic maps that document the Indigenous languages of Australia for they were among the coastal groups that felt the full brunt of colonial expansion. Wilangarra, including men, women and children, were massacred. Most were

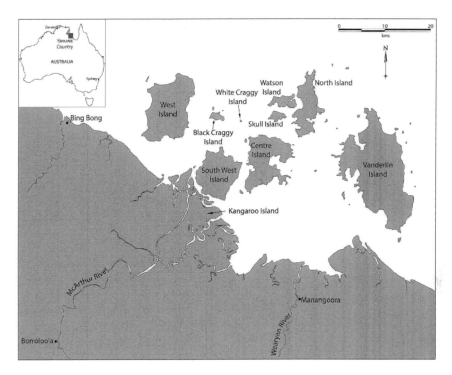

Figure 2.0 Yanyuwa Country, southwest Gulf of Carpentaria, Northern Territory, Australia.

killed by police and quasi-police groups, and by drovers and station workers involved in the cattle droves of that era (Roberts 2005: 66). This 'frontier violence', so powerfully conveyed by Roberts (2005) was driven, in 1881, by a "massive pastoral boom [which] commenced in the top half of the Northern Territory" (Roberts 2009): "By the end of the year [1881] the entire Gulf district … had been leased to just 14 landholders, all but two of whom were wealthy businessmen and investors from the eastern colonies". Roberts (2009) recounts the rapid sequence of killing that came with the expansion of cattle leases. Landholders had three years to comply to the minimum stocking rate, and it was during this time, that Roberts (2009) reports, "at least 600 men, women, children and babies, or about one-sixth of the population, were killed in the Gulf Country to 1910. The death toll could easily be as high as seven or eight hundred".

There was no regard for Wilangarra life, nor recognition of existing rights to and order in place. There were no explanations and no consultations for the taking of lands and waters (Roberts 2005, 2009). Despite the events of this period being horrific, settlers reported them to authorities without caution. These accounts highlight the annihilation of Indigenous lives. Roberts

(2005: 2–5), in his account of this history – a history that remains largely hidden – describes the effects of cultural wounding:

> In just four years, the Aboriginal population of at least 4000, composed of 15 tribes or language groups, was dispossessed of every inch of land. Profound cultural and emotional consequences flowed from this, made more acute by the complex spiritual link that traditional Aboriginal people have with their country. Compared with more than 35,000 years of occupation, four years must have seemed like the blink of an eye.

Amid these cruel encounters, country lost its most necessary companion – its human kin. It is believed that in the aftermath of this violence that there were no Wilangarra-speakers left, and while a few Yanyuwa elders, many now passed away, had Wilangarra ancestors, the group is no longer considered a landowning presence in the region (Baker 1990: 27). Still today, Yanyuwa elders will describe the western part of their country as "mix up country, little bit Wilangarra, little bit Yanyuwa, but close up [which refers to the linguistic and cultural traits shared between the two groups in the past]" (Kearney unpublished fieldnotes 2003; see also Bradley 1997: 58).

Yanyuwa experiences of frontier violence were far from benign. There is a living archive of oral testimony that recounts the violence of settler colonialism (see Baker 1990; Roberts 2005: 189–199). Their homelands have been taken, their children forcibly removed as part of the well-documented Stolen Generations (see Read 1999), their lives institutionalised and constrained by the policies and governance of the settler colony and the Australian state. Lives have been made extremely hard and, across the generations, Yanyuwa have repeatedly had to fight for the small right to live their cultural lives, raise their children according to their Law and care for the country that is such a vital part of their existence. Wounding has come in the form of lost rights and restricted access to homelands, loss of lives, near-complete linguistic death, poverty and experiences of deep racism (Kearney 2014: 109). Yet they stand resolute as the Indigenous owners of land and sea territories that are encompassed by the delta regions of the McArthur River and the saltwater limits of the McArthur and Wearyan River, the Bing Bong area and the Sir Edward Pellew Islands (see Figures 2.0 and 2.1).

The offshore islands and immediate coastal areas represent the foundation of Yanyuwa thought and existence. By emphasising themselves as sea people, as opposed to mainland people, and establishing a "proper" relationship to country, Yanyuwa activate a human:country relationship that is distinct (see Bradley 2008). This relationship is made vital, managed and expressed in a variety of ways and manifests itself in language, kinship, patterns of settlement and subsistence, song, ceremony, ancestral and social narrative (Bradley 1998, 2008, 2010). Each of these features of Yanyuwa culture and the desire for cultural distinctiveness are echoed in the expressions *li-Anthawirriyara* – people of the sea, and *Yanyuwangala* – the essence of being Yanyuwa. Country

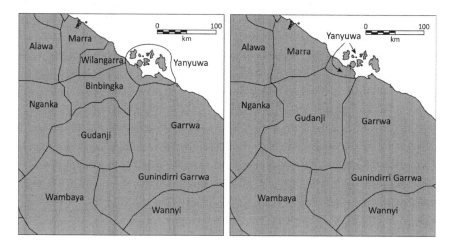

Figure 2.1 Approximate linguistic boundaries in the southwest Gulf of Carpentaria, Northern Territory, Australia. The map on the left shows pre-contact and early post-contact linguistic boundaries, and on the right, present-day linguistic boundaries.

consists of more than the physicality of the land and sea, and incorporates the wider themes of the elements, animals and people, relationships, obligations and interactions that accompany people's relationships with place.

Yanyuwa country as emotional geography

While the name Bing Bong is used throughout this chapter, it must be mentioned that this is not an Indigenous place name, and is not derived of either the Yanyuwa or Wilangarra languages. Rather it is a name that was given to this place by non-Indigenous pastoralists. The name, which is now widely used by the region's Indigenous and non-Indigenous residents, is said to come from a famous racehorse. Yet there are few who are sure about this, and there is little documentation to say how or when this name came into common usage. It appears on contemporary and historical maps, yet this is the case for many non-Indigenous place names throughout the region and all of Australia, where Indigenous place names have been obliterated by cartographic colonial inscription. This is one part of the disordering intent that has happened throughout the region, with the attempted erasure of toponymic distinctiveness. But it is also a testament to how strongly place names hold on, especially when linked to the bodies and actions of ancestral beings.

Yanyuwa know the place names for this region, and while they might make reference to this area as Bing Bong more broadly, this is not evidence of them accepting erasure nor forgetting the sedimentary layers that are contained in these places, and iterated through their names. These remain, and as one Yanyuwa elder affirmed in 2003: "It's Aboriginal land, white man lease over the top, but you can't get him under ground, you can't get that one, you can't take him name or story, you can't kill him" (Dinny McDinny Nyilba in Kearney unpublished fieldnotes 2002).

Yanyuwa have sophisticated ways of negotiating the colonial presences that have attempted to write themselves into country through naming. There is no more vivid an articulation of this than their production of the Yanyuwa Atlas, *Forget About Flinders: A Yanyuwa Atlas of the South West Gulf of Carpentaria* (Yanyuwa Families et al. 2003). A response to the widespread renaming of their country by Matthew Flinders, a British explorer who sailed around and mapped the Gulf of Carpentaria, Yanyuwa introduce the volume with the following statement:

> This atlas is about the Yanyuwa story of the country that Flinders travelled around. Nobody wants to forget about Flinders because he was a very remarkable person, but it is important to remember that everywhere Flinders went there were people, Yanyuwa people, and every name that he gave to the islands and the sea, already had a name, a Yanyuwa name. So really this book is about forgetting Flinders for a while so the Yanyuwa story of the islands [country] can be rediscovered.
>
> (Yanyuwa Families et al. 2003: i)

The name Bing Bong appears to have etched itself into local orthography by 1964, when the Northern Territory government, granting pastoral landholdings for a huge portion of Yanyuwa homelands, signed away the Bing Bong Pastoral Lease. Very few, if any western maps feature the Indigenous toponymic details of this region, which eclipses the singular place name of Bing Bong and instead instates a whole series of names that give rich detail and meaning to the coastal area. Each name pertains to the actions of ancestral beings, as well as the events that created country, clan distinctions, and the Law that governs everyday and ceremonial life. Stating the names Makukula, Mawuli, Arrinyanda, Wimanda and Wurrulwiji invokes an emotional geography across generations, and also triggers recollection of my own relationship with Yanyuwa families and their homelands over the past 15 years. These places at Bing Bong were some of the first locations I was taken to by Yanyuwa as we embarked on collaborative ethnography in 2001. I have always travelled to the area with community members, most often senior women and their grandchildren. There are prohibitions against disassociating oneself from networks of kin, which position the individual within a matrix of relations and meaningful interactions. It would be very strange for a Yanyuwa person to travel alone on any part of country, and during my time there I have been subject to such conventions as well.

In fact, I have never felt inclined to travel alone across Yanyuwa country. On the few occasions when I did venture off independently for a short walk, doing the classic 'whitefella' gesture of 'needing some space' or 'clearing my head' (being largely unaccustomed to the expectations and demands of extended kinship), even if only several metres away from the group, I was quickly reminded of Yanyuwa Law and the sentiency that resides in place. Communicative events, even for my novice mind, were such that whenever alone on country, I encountered something that either startled me, scared me or baffled me. I recall moments of becoming suddenly engulfed by networks of rock pillars and overgrown vegetation, of forgetting my pathway back, to stumbling across groups of animals or their tracks, disturbing their midday rest and causing a commotion that could be often terrifying. There have been instances where I am acutely aware of hearing my name called out on country (I have even responded), hearing motor vehicles or boats approaching, and then finding no one had been calling out or traveling nearby. A sense of disorientation can quickly descend when on country alone and with knowledge shared to me by Yanyuwa, I have become aware of co-presences and other agents existing on and in country. In their entirety, my understandings of place have been deeply complicated and cross-culturalised over the past 15 years of learning about Yanyuwa country and co-presences in place.

Unsettling encounters on country, when reported back later to my Yanyuwa hosts and extended social family were always interpreted as reminders of why not to 'go it alone' on country, as communicative events attributable to the old people, who know, see and hear everything. I have been warned that the old people might come and choke a visitor, they might role you up in your swag (camp bed) to scare you, they might entice you away from the group to satiate their loneliness, or trick you to satisfy their cheekiness. I trust Yanyuwa in the advice that these ancestors are always watching and aware of what goes on in place and, in fact, that place itself is always watching. While this principle applies to all, in that the old people and ancestors can see all – including Yanyuwa and non-Yanyuwa – there is specificity to the interaction when one is not from a particular part of country. In sum, the risk is greater if you do not have kinship with place. This means that kinship is not solely the realm of the human, or biological relations, but is a governing system of Law and order of which place and the old people are also keenly aware.

The power of country is such that if it does not know you, if you are not Yanyuwa, not its kin, and it cannot recognise your smell, stride, mannerism, language and essence, then the ancestors may reject you in the most harsh of ways through the causing of sickness, death, wild misfortune or disorientation. Because of this, the need to 'call out to country', the singing and spoken oration that is performed as people arrive at places becomes paramount. With this practice, a call is made by senior persons, either elders or *jungkayi* (Yanyuwa term for guardian, which applies to one's mother's country), to the old people and the co-presences in country. It may take place as a boat approaches the coastline, as a group of people step ashore onto an island, orated from the shallows all the way up the beach. It might be sung as an approach is made to an old campsite or shelter known to harbour rock art or log coffin burials. It is necessary when approaching country that has not been visited in some time, with the need to sing out becoming stronger as kin are absent for longer. Even for country and places that are visited often, the need to sing out and announce presence and the substance of that presence is essential.

Langton (2002: 262) writes of the need to 'sing out' to country on arrival and approach. Singing out to country, and speaking to ancestors, is described as an act of "looking after country" (Langton 2002: 262). She details the decision to do so by George Musgrave, of Thaypan country, Cape York, northern Australia. The obligation to sing out comes from the knowledge that it is the ancestors who are the "keepers of life forces – life's fecundity" (Langton 2002: 262). The singing out is done not only for the sake of the Indigenous owners but particularly for the sake of others who might travel with them (Langton 2002: 262). "The task that falls to Elders in these circumstances is to protect those who are traveling with them from the dangerous spiritual forces emanating from emplaced spiritual beings" (Langton 2002: 262). When this address is made, it signals that the Indigenous place kin, and the ancestral beings that constitute place, are on "close speaking terms". Langton (2002: 262) describes an oration that was shared with her: "Look old thing, it's just me. Alpulkal. You know me. I've got some friends with me. Don't hurt them." While the translation is provided, Langton (2002: 262) notes that George (the narrator) "must speak this in his own language, the language of his ancestor, to reassure it that there is no stranger but someone who shares ancestry and language". Old people and "deceased ancestors … never depart from the landscape. The land is full of spiritual presences which are the ancestors of traditional owners, usually grandparents. That is, they were people known to the speaking subject in their own lifetimes, and they must be spoken to in the appropriate language" (Langton 2002: 262). "If one does not 'sing out,' there are dire consequences. Engaging with a place thus requires engaging with the spiritual presences therein" (Langton 2002: 262).

Indigenous persons undertake actions of this kind across Australia. Many reiterate the same principle of familiarising the ancestral beings in country with unknown persons, as well as instating kinship that already exists. In the case of the latter, the act identifies the specificity of kinship between people and place while also signifying the responsibility that persons have for any stranger they might bring onto country (Rose 1992: 109). The stranger, in acknowledging this lack of kinship, assents to the fact of power in place, that others are owners, that co-presences prevail and that they themselves do not have the knowledge or know the order of place (Rose 1992: 109). The specific intonations and linguistic references Indigenous kin may use in calling out to country may vary from group to group, but the message is the same; a recognition is given to the multifarious agents in and of place to whom the person is kin. And if not kin, their existence is explained and presence justified. Rose (1992: 109), notes that for Bilinara and Mudbura people in northern Australia, and as described by Hector Wartpiyarri, calling out or singing out to country is a practice whereby "You [the owner] talk [to the country]; 'I got *ngumpun* [Aboriginal/human being] here from long way. I don't want you to hurt him, he's *jimari* [mate], he's all right. [You] take him, give him water, wet him head. Now he can walk every way [anywhere in that country]. That's *kamariyu* – from long way."

According to Rose (1992: 109), "people explained that once a person has been introduced to the country through this means the country knows the person's smell. Without this introduction, strangers are at risk – the water may drown them, or they may become sick and die." In this case, Rose is referring to the Bilinara and Mudbura practice of taking people to parts of country where there are bodies of water that are known to hold spirit beings, or that are the embodiment of ancestral beings (for example, freshwater lagoons, rivers and wells). Through ethnography, she recalls, "Jimmy Manngayarri said that it is the Dreamings [ancestral/creation beings] in country who actually smell and identify people's sweat; 'Dreams [ancestral/creation beings] can smell other people. After you water him [anoint a person in local waters], that Dream knows him'" (Rose 1992: 109). Some groups are in the practice of wiping the sweat from their armpits onto the earth to convey the essence of the person, a sensory encounter in which smell is taken to reveal the interior and exterior qualities of a person who wishes place to know their character and intent.

Through its sentiency, place is capable of detecting this smell and responding to it (see Kearney 2009a). Singing out to country is a vital step in opening the communicative pathways between people and place, and when you are a visitor to country, failure to do so is a highly risky undertaking. This is not a risk I have ever dared take; instead I seek the company of Yanyuwa as my most necessary companions when traveling on country. It is Yanyuwa who are capable of communing with place, introducing me and ensuring safe passage. They are the ones who speak a language that place understands and there has never been a clearer expression of communication between multifarious agents than that displayed by Yanyuwa calling out to country.

For Yanyuwa, calling out to country is akin to saying, "Old People, I recognise you and your pervasive Law, I come here as someone who belongs to this place. You know me, as I am your kin. You will accept me here. With me I bring a visitor, who is ok for this place. I take responsibility for them and they will follow the Law." Orations may be formal in their language and delivery, while others will have a sense of the everyday in their styling. It might be that a woman goes fishing, and speaks to the old people and country, by saying, 'jaba arlku!' which means 'give me fish' (from John Bradley unpublished fieldnotes, pers. comm., 25 November 2015). Place and the old people then respond by giving fish. If place and the ancestors do not meet this request, then another type of communicative event has taken place, one that must then be read and interpreted accordingly.

This aligns with Povinelli's earlier work on 'sign reading', in which reading signs – such as absence of species and food; anomalies of species shape, behaviour and location; shifts in elements and natural phenomena – contained in country is associated with risk management, reciprocity, social action, politics, ecological probability and happenstance (Povinelli 1993: 684). A more formal oration to country will take a different form, drawing more closely on the nuances of kinship and obligation. This is a powerful declarative act, which instates relationships, as illustrated by the words of Yanyuwa elder Johnson Timothy, as he approached Liwarrangka, an extremely powerful place on South West Island in the Gulf of Carpentaria. Only men may visit this place, in the company of *jungkayi*,[1] for a high level of Yanyuwa Law governs it. On this day, in 1984 on approaching Liwarrangka he orated:

> Wayi! Listen to me! I am standing here a guardian for this place do not pretend not to know me, I am here, I have been given eyes to see this country, my younger brother also stands with me, he too has been given eyes to see. Listen to me! Standing with me is a child of this country, one whose spiritual essence is derived from here, also there is one who stands in a mother's mother's relationship to this country: Do not strike him! Do not make him fall! I am here a guardian, do not pretend not to know me!
>
> (John Bradley unpublished fieldnotes, 1984, in translation,
> pers. comm., 25 November 2015)

Included below are a series of Yanyuwa orations and calls to country, each translated from Yanyuwa into English, and conveying kinship between people and place. Each involves a Yanyuwa person and a Yanyuwa place. These illustrate the communicative patterns that exist between people and place, revealing the intimacy between the two as kin. This brings forth the emotional geography at play here, characterised by a deep love and intimacy for place kin.

Context

In 2000, Jerry Brown Ngarnawakajarra was travelling to Bing Bong with anthropologist John Bradley. Upon arriving at Bing Bong, somewhere he had not been for some time, he offered an oration as he stood halfway between the pastoral homestead and the mining port facility, at Wurrulwiji.

Oration

Ah dear me, I am crying. I have tears in my eyes, I am rubbing my eyes. I have not been to this place for a long time. Listen to me! I am here a kinsman for this country. I grew up with the old men for this country, they gave me eyes to see the Law for this country. Listen to me! I am standing here alone, the old men have all died. I am here do not pretend not to know me! Listen to me! You are my kin, your essence is the same as me, do not pretend not to know me. This country is blocked, it is shut up, you have entangled yourself, what has happened here? I am blind in this country, you are shutting up, closing in, where is the open country, when I was young I could see a long way, from here to the sea. Oh you poor thing, the old people have all died … no fucking good altogether (Bradley pers. comm., 25 November 2015).

Context

In 1999, 79-year-old Hilda Muir returned to Yanyuwa country after years of separation, caused by her being part of the Stolen Generations of forcibly removed Aboriginal children. At the urging of her extended networks of Yanyuwa kin, it was decided that she should be "reintroduced to her mother's country" through oration and singing out to the ancestors (Bradley in Muir 2004: 142) in adherence with Yanyuwa Law. On this return, an oration was made.

Oration

Here she is, one whose mother came from this place, her mother was a-Manankurrmara and this is her child standing with me. She has been away for a long time but now she has come back and is crying for this country. Please do not ignore her, she was away but now has returned. She is truly a kinswoman to this place (Bradley in Muir 2004: 144).

Context

After the recent death of Jemima's husband, her extended family decided to return to Kangaroo Island. Her husband, a much-loved community leader and man of great intellect, had died on the island. The sudden nature of his death caused a profound pain within the community. People grappled with the possibility that his own country had killed him. On the return to the island, Dinah Norman a-Marrngawi sat by the fire, watching carefully the long-necked turtles that were cooking in the ashes. The sun had not yet risen but already it was warm. She reached for her digging stick and turned the long-necked turtles, and recovered them with the cooking ashes. She then stood up, crying out loudly (Bradley 1997: 182–183).

Oration

Hey what is happening here? I am here a child of this country, my spirit is from here. I am a kinsperson to this country. What is happening? This long-necked turtle is small, too small! Listen to me! Do not pretend not to know me! She then sat down looked across to her son who was watching a group of galahs gathered on the nearby plain. She said matter-of-factly, "She is a kinswoman to this country, she carries it, she lifts it up when we are not here" (from Bradley 2012: 28; 1997: 182–183).

Context

It is 1999, a group of Yanyuwa are traveling across their country proofing land claim evidence (the process of establishing fact by reviewing claimant testimony) for an up-coming land rights claim, under the Aboriginal Land Rights Act (Northern Territory) 1976. As they travel up the McArthur River, the boat hits a cross current and is dipped down into the trough of a wave. The crest of the wave comes over the boat and wets everyone. Dinah Norman a-Marrngawi, who calls the wave mother, and who is *jungkayi* for the Wave Ancestor at another Yanyuwa place, Muluwa, responds immediately.

Oration

"Hey! Here it is, I am your child, do not pretend not to know me. Do not throw your spit at me!" (Bradley pers. comm. November 2015).

Over the past fifteen years, while visiting Bing Bong with Yanyuwa families, we have fished at Makukula, told stories and sat and listened to country at Arrinyanda and Wurrulwiji, hunted for freshwater turtles at Wimanda, carried out interviews underneath coastal pines, searched for bush medicine, sung and gossiped about life in town. These visits, however, have become increasingly rare. They have never been easy in my time, and have always required permission from the pastoral station managers to unlock their gate and tie up the dogs so as to let us pass (see Figure 2.2). They became harder as more areas were fenced off or left to choke up with weeds, as pastoral station managers changed and became less forthcoming in granting access to Aboriginal land that lies to the far side of Bing Bong Station. They were to become tricky as coastal dredging occurred and land disappeared in order to accommodate the expansion of a longstanding mine ore loading facility (belonging today to Glencore, formerly Xstrata). Then in 2013/14 another mining company (Western Desert Resources) constructed a major access road through the area and built a large conveyer belt to transport iron ore concentrate across the vast coastal area of Bing Bong (see Figures 2.3, 2.4 and 2.5).

During a visit, ahead of writing this chapter, in April 2015, we found Bing Bong stained red with the dust of iron ore, dispersed into the air from the uncovered conveyer belt transporting the concentrate. Left uncovered, the concentrate billowed up and was spread by wind currents. The effect was a ghastly shade of reddish-pink on all living things across the Bing Bong area. The yellow-crested cockatoos and goannas were stained pink, the freshwater lagoons covered with a dry powdery coating, the trees stained red and desiccated by interruptions to local water sources. Combined with the visual effect of debris mounds, heaps of old tyres, bent metal, used and broken machinery, barbed wire fencing and signage, the whole area felt, to me, unable to breathe.

In October 2015, I returned to Bing Bong and found the conveyer belt inactive, the company having gone bankrupt due to global falls in iron ore prices. There was no sign of remediation, and there are still no plans to

Figure 2.2 "They put the gates up everywhere now", gate at entrance to Bing Bong Pastoral Station, 2003 (Photo source: Amanda Kearney).

remove this mess from country. It stands there like an ugly broken down metal giant, a testament to the fragility of this mining boom and the greed of neoliberalism. Country is so dry and so hot you can barely stand five minutes at the gate, where signage points its finger and antagonistically declares, 'Mining lease – no entry', 'Strictly no access permitted', and 'Unauthorised entry is an offence under Australian Maritime Security Regulations' (see Figures 2.6 and 2.7). The beach and coastal margins that were Arrinyanda are now gone in their entirety, having been dredged into oblivion; Wurrulwiji has been dried out and tidal patterns heavily affected as dredging has shifted the coastal margins. There are no kangaroos to be seen, their tracks replaced by those of feral cats, pigs and cattle, and the Winter Rain Dreaming at Wimanda is filthy with toxins, cane toads and smells like the rot of disordered country.

On this last visit I stood atop the Bing Bong lookout – a metal scaffolding-type construction perched amid the mining port facility and the conveyer belt – 'generously' erected by the mining company (Xstrata, now Glencore) – and took in an unfamiliar scene. It was a sight of diminished presences, vacant and empty, damaged and harmed. In fact there is really nothing to see from this lookout other than the mine's loading facility and the conveyer belt, so successful has the land clearing and destruction of country been. As if incapable of keeping up with their own devastation, the mining company

Figure 2.3 Bing Bong, Wurrulwiji in 2010, before the construction of a large conveyer belt to transport iron ore for Western Desert Resources (Photo
source: Amanda Kearney).

once placed a panorama photograph on the lookout railing. It is mounted on
the edge of the rail, and the visitor is encouraged to look at the image, look
to the landscape and get a sense of what they are seeing before them. Yet
the two no longer match up. The photograph, most likely placed there in the
late 1990s, is a relic of what this series of places once looked like. The image
shows the coastal margins of Bing Bong, the noticeable features of Mawuli,
Arrinyanda and Wurrulwiji, the contours of country captured in a fading
black-and-white photographic panorama. But in this photo, their presence is
still substantial, the area is forested with mangroves, transected only by small
tracks down to the beachside. It prompts a double take, from the photo, to
the scene and back again. They look nothing alike. Chunks are missing, the
land is gone, trees are gone, roads have been laid down and fences put up.
There is no way of concealing that these places are either missing or entirely
rearranged (see Figure 2.8).

    As I stand on this lookout I ask myself, where have the old people gone,
where did the ancestors who reside in country seek refuge, what of all the
animals that share in these places? Where on earth did they all go, and can a
place still exist if its physical presence is either gone or so drastically modified

Figure 2.4 Bing Bong, Wurrulwiji in 2014, after the introduction of infrastructure to support iron ore transportation from the Gulf of Carpentaria (Photo source: Amanda Kearney).

as to resemble nothing of what it was before (see Figures 2.2–2.5)? Many Yanyuwa have stopped visiting Bing Bong. Some gave up a time ago, and prefer to remember place and country as it was, although they know it's not like that anymore. They hear the stories back in town, and are told of how 'buggered up' country has become. In October 2015, elder, Dinah Norman a-Marrngawi spoke about Bing Bong. She first took me there in 2001, as I began my doctoral research. Bing Bong has been on her mind a lot lately, through talk of its decline and through the death of her sister, a senior woman whose *ardirri* (spirit child) came from this area. All of this has promoted her to orate, and ask,

> – *kanu-wingka*
> *nganinya nguthundiya*
> *walkurriji Makakula*
> *wayi ki-awarala?*

> – We used to go
> this way northwards
> to camp at Makukula
> I wonder how that country is now?

Figure 2.5 Close-up of operations at Bing Bong, Wurrulwiji, 2014 (Photo source: Amanda Kearney).

Figure 2.6 Signage, Bing Bong (Photo source: Amanda Kearney).

Figure 2.7 Signage, Bing Bong (Photo source: Amanda Kearney).

Figure 2.8 Aerial photograph of Bing Bong, highlighting the designification efforts of agents of harm, across the wider area (Photo source: Google Maps).

Dinah's comments, while appearing subtle, echo a sense of grave concern for these places. She is at once negotiating her responsibility for Makukula and its surrounds, entering into a dialogue between herself and the old people. The question is rhetorical in many respects, and through her wondering, she instates a relationship to place and the ancestors, as an act of caring, an empathic moment of reflection. Nobody offered an answer to her question, understanding that it is not one to be answered. It is not unlike the Yanyuwa understanding of 'management', a principle that has gained attention in an era of expanding natural resource management within the region, in that it echoes an inherent Indigenous epistemology contained in caring for country (see Baker et al. 2001; Bradley 2001).

For Yanyuwa, 'management', is best aligned with the axiology of caring for country, recognising and enacting kinship. Ontologically this maps across a number of actions, from standing on country, to thinking and talking about country, of articulating the relationships between self, group and country through hunting, singing, visiting, remembering and practising ceremony. These are all forms of active management, whereby management is a praxis constituted dually by action and emotional engagement. This is management in the fullest sense, more expansive than other management acts such as weed eradication, feral animal trapping and ghost net collection (see Kearney and Bradley 2015: 171). Dinah is fully aware that Makukula and all of the surrounding places have been deeply wounded. She is cognisant of what has caused this and searches through the epistemological structures provided for by Yanyuwa Law to assess the implications of this, adopting the appropriate emotional response to its gravity. As kin, she must reflect on country and dedicate part of her emotional reservoir to feeling for these places and the ancestral beings they contain. This is an emotional geography that runs deep, and opens a person up to emotional states such as grief, shame, fear and longing. Her oration also evinces something of the history in which harms have been done and Yanyuwa have been unable to stop them.

The emotional geography of Yanyuwa country emanates out from '*awara*', '*ngalki*' and '*narnu yuwa*'. These are Yanyuwa terms saturated with linguistic profundity. While there is difficulty in translating their meanings into simple English, attempts have been made to clarify their significance and potency as part of an Indigenous pedagogy of place (see Bradley et al. 1992; Bradley with Yanyuwa Families 2016). Yanyuwa use the term '*awara*' to refer to place and country; but its more expansive meaning takes in the vast elements of ground, earth, land, camp, soil, sea, sand bar, mud bank, reef, situation and home (Bradley et al. 1992; also Bradley 2008: 21).

Awara blurs the line between a contained sense of physical space, and the more intangible qualities of country, as made up of essences, experiences and internal or ancestral workings. People speak of *manhantharra awara* as the act of embracing country, or *wukanyinjawu ki-awarawu*, speaking to 'country', today done most often through the oration, 'hello my countrymen'. *Awara* takes in the whole sense of multifarious agents in and on country (see Bradley and Yanyuwa Families 2007). *Ngalki* also does this, by denoting the 'essence' of all living things. It is what gives a place its character, a flower its perfume, an animal its movement and an individual the scent of their body sweat (Bradley 1997: 150). *Ngalki* is held to be like 'skin', it is what gives shape and substance to all living things on Yanyuwa country, including the place world (Kirton and Timothy 1977: 321).

There is specificity to place as a presence, which holds all the interacting elements of country, from the human to non-human. This specificity exists because of *narnu yuwa*, that is the Law that governs Yanyuwa country and life. *Narnu yuwa* comes from the *yijan*, a time when the ancestors travelled across the land and sea forming human and non-human life, kinship, places, place names and imbuing the place world with order, identity and potency (Bradley et al. 1992; Bradley with Yanyuwa Families 2016). The relational logic of place is tracked through its being *awara*, given existence and substance by *narnu yuwa*, which comes from the *yijan*. Human and non-human life are implicated in each part of this sequence, thus they are also part of *awara*, subject to *narnu yuwa* and were created by *yijan* (Bradley 2003: 5). These aspects of life are indivisible and absolute. What happens in one realm is nested in another. Law is the centre of the Yanyuwa universe, hence its obliteration at Bing Bong and elsewhere on country is unforgiveable and the source of much sadness and human suffering.

The nested ecology of Yanyuwa country is such that emotional dispositions, such as loneliness, sadness, joy and elation, can be experienced by place, just as people experience them. Place as an agent and sentient co-presence has its own emotional capacity. It can feel and respond to other presences and in doing so communicate the health of country, the strength of kinship ties and interactions that it is experiencing, but also the disorder caused by moral disengagement and axiological crises of ambivalence and violence. It is also capable of communicating an emotional response to the absence that is felt when kin have not been present or active in their roles and responsibilities. These expressive dispositions are referred to as 'country being up high' and 'country being down low' and are observable through place's expression of health or ill health. People refer to country as 'full of energy, strong and glowing, brilliant and bright' or conversely country being 'run down, weak, and lonely' (Bradley 2008: 22). When country is strong, the orations are akin to a conversation, a gentle dialogue between kin, acknowledging the beauty of country and the sense that all is well and capable of being seen and heard. Such dialogues are said to lift the spirit of people and place, as is the case in the following witnessing of country in all its strength, made by Dinny McDinny Nyilba (in Bradley unpublished fieldnotes, 1988, pers. comm. November 2015) on an occasion when camping on country with extended family.

Ngabuji [father's mother, he is addressing the country] here I am, I have returned. It is sometime since I have been here, I was away but I have returned, ngabuji. We have always moved through this country, my mother's and mother's brothers who came from Kalalakinda [place located on the banks of the Foelsche River] they would come here. We always used to sit together, we never used to go away from each other, we used to help each other in times of trouble like in fights. But all my mother's family have died now, some also went away to Doomadgee. But before this they always sat down together all the family. They used to stay on their country there in the east on the Foelsche River and the Wearyan River, they never moved away.

Hey! All of you with me! Wake up! This country will not know you if you do not talk, if you do not move, why are you all asleep! This country will close your eyes and ears! Wake up!

I am here talking alone, ngabuji we have come a long way. We have slept here and then we will go east, my country is there in the east, it is good country, a country of plains and lagoons, but I am here on my ngabuji country, I am here talking alone, this is the way old people would talk to the country, that country would listen and know the people when they moved hunting or going for ceremony, it would know.

Oh dear me I think and I am sad, when the white people came we left this country, this country where we all grew up and they went up to live at Borroloola. Welfare rounded us up, brought us in, we did not want to leave our country, but welfare rounded us up like a bullock and all of those dear old people died, they left their country from here and into the east where they lived on dugong and cycad food and fish that they caught in fishtraps.

Hey! My kinswoman the crow is calling out, yes there you are my most senior paternal grandfather's sister, I hear you, you are waking up the country, this is the way it should be.

Ngabuji we lived in this country, we hunted, we danced ceremonies, we settled disputes. We would fight intensely. It was a hard life, we fought over ceremonies, run away lovers, children, wives, husbands, dead people's bones, I remember these times … aah the crow my kinswoman is still talking, I am here! I am here! I have not forgotten. These young people still sleep, they do not know this country yet, not yet, they might learn, wake up! This place has to hear you! This place has to see you move! Get up! Ngabuji I am finished now, I am still here, I still remember this place and all the country into the east.

Such is the richness of a sensuous geography, that witnessing both beauty and decline in place, and across country is a vital part of the communicative relationship. A more commonly encountered oration since the arrival of a settler colonial presences and their destructive practices on Yanyuwa country are those that carry a tone of sadness, and even shame; emotional dispositions directly related to the intimacy that binds people and place.[2]

This country is a little bit low, it is a sad place, they have all gone, all the old people who hunted and sang and danced many ceremonies here. The children have gone, those who used to laugh and play here, they have gone, the land is silent. The women have gone, those who used to hunt and dance they have gone. There is only me here, this country is weak. In times when I was a child this country would move with people … food was everywhere in abundance, there is nothing now, the old people have gone the country is weak, we cannot hear it.

(Johnson Timothy Rakawurlma 1990, in Bradley 2006: 178)

Determinations of place decline, and of country being 'low down' have increased over the past few decades. This is part of a long-term trajectory of harm that came with settler colonialism and cultural wounding. Exile and physical estrangement mark a large part of Yanyuwa experiences of place since the arrival of settlers, an all-too-familiar scenario in colonial settings. The power of place, however, continues to be asserted by many Yanyuwa, and for elders and mid generations more closely educated in the details of Law, this is articulated through the expressions and ontologies of *wirrimalaru* and *wirriwangkuma*. These denote the essence and order of country and signal information on the nature and power of place. It is through an exploration of what these meanings contain that we move further into the wounding experience and what is really at stake when axiological crises take hold and cause the assault of place.

When speaking of country, Yanyuwa use the term *wirrimalaru* most frequently. People refer to certain locations using the descriptor, 'big place'; to denote a location that possesses *wirrimalaru*. *Wirrimalaru* means something of great importance or power. Dreamings and their related ceremonies are said to have *wirrimalaru*. A place considered to possess *wirrimalaru* embodies high position, spiritual, symbolic and political power (Bradley 1988: 32). As elder Gordon Lansen Milyindirri (Kearney 2009b: 190) once reflected, *wirrimalaru* is "[j]ust like a prime minister, it's big boss country with a big name. No one can touch im, you gotta look after him all time. *Wirrimalaru* country always strong, forever … kids [ardirri – spirit children] been found here, people come together [in these big places], you know!" In 1990 Baker recorded a number of locations on Yanyuwa country referred to as 'big places' (Baker 1990: 41). The importance of each location is linked to the ancestors, people's patterns of visitation, presence of rich resources, fresh water and the number of conceptions and births found at these places.

The places that constitute Bing Bong are *wirrimalaru awara* because they are the ancestral embodiment of *a-Buburna* – the Black Nosed Python and *Murnnyi* – the Winter Rain Dreaming, ancestor for the Wuyaliya clan.[3] There are songlines associated with this Dreaming that map country, placing locations in relation to one another. Bing Bong exists as one part of the

Dreaming's travels; it was the Black-Nosed Python who travelled into and made the country west of Bing Bong, she coiled up there and became that place, but as she woke she looked to the north and let out her great expiration that would pass over country to create the Winter Rain Dreaming at Wimanda (Yanyuwa Families et al. 2003: 279). She then continued her journey east where she was set to create other places. Passing in close proximity to Bing Bong are also the *li-Maramaranja* – Dugong Hunters (Yanyuwa Families et al. 2003: 277) and there is a turtle increase site along the coastal margins of Bing Bong and much of the coastal area of Bing Bong and the offshore islands are associated with the path of the *Wundanyuka* – Green Turtle Dreaming. Various rock formations along the Dreaming path are the shell, internal organs and segments of meat, which the hunters of sea turtle most highly regard.

The spirit children for many present-day elders were found at Makukula, Wimanda, Arrinyanda, Wurrulwiji and Mawuli. Many others were born here during the 1940s, a period in which ceremonial life was active and strong in the region (Yanyuwa Families et al. 2003: 279). The combined effect of these presences and experiences was the enrichment of place and lifting up of country. The strength of the old people, and thus strength of place, came from the ontological commitment to kinship in action. Enhanced further throughout the mid-1970s and 1980s, country was routinely lifted up and strengthened by the presence of families who would camp at Wurrulwiji, Arrinyanda and Makukula for extended periods during the school holidays. In the 1990s, however, these regular visitations began to stop, as a direct result of mining interests commencing operations in the region and plans to develop the coastal area as an ore loading facility.

Like *wirrimalaru*, *wirriwangkuma* gives expression to the identity of place, as well as people's relationship with place. It is a quality recognised in places that possess a long history of communal gathering and intergroup activity, places that offered kin the space and inspiration for substantial ceremonial activity. *Wirriwangkuma* is applied to only three parts of Yanyuwa country, one being the coastal area of Bing Bong. As Bing Bong marks the only point on Yanyuwa country with ready access to the sea, it provides richness in marine resources, including dugong and sea turtle. It is rich in foods, but only because the ancestors make it so. Recognition of *wirrimalaru* and *walkurra awara* (big country) carries with it a range of responsibilities, moral obligations, social codes and expectations, all of which must be upheld by those standing in a clan-based relationship to these parts of country. By enacting this kinship, both through physical acts and cognitive processes Yanyuwa uphold their end of the relational bargain. As they say, "we Yanyuwa people, we cannot forget about this country, we are continually thinking about it, this country that was truly for our ancestors, we are thinking about them all the time" (translated in Bradley 1988: 47). That this ethos has struggled under the weight of colonial invasion and more recent mining expansion and its associated place annihilation is abundantly clear.

What is also clear is that the harms done to place have been achievable because of a disregard for the human life nested within. The intent to culturally wound and do harm was explicit in the killing of Indigenous people throughout the region, marking the viciousness of frontier encounters (Roberts 2005), yet has been veiled (only thinly) in more recent decades with the arrival of neoliberal agents. What follows is an unveiling of this destructive presence for what it is, tracking the sequence of place harm as a seamless operation, from colonial killings and dispossession to the arrival of the great mining giant that has cut country so deep that it may never recover.

Destruction and social disorder: the arrival of colonising agents

At present, most Yanyuwa live in or near the township of Borroloola or have chosen to reside at outstations throughout the surrounding area (Baker 1999: 10). The township was established in 1885 and in the ensuing years would come to be known as a frontier outpost, regarded for its wild characters. It is located inland of Yanyuwa country, some distance from the coast and saltwater influences of the major river systems. From the mid-1800s, Yanyuwa relationships to country were rearranged amid a suite of invasions and impositions, which called on them to resist, accommodate and entangle with a non-Indigenous presence in the region. The Borroloola police station was set up in 1886 in an attempt to curtail the lawlessness that had come into this region, and, according to Avery and McLaughlin (1977: 3), journals left by police officers describe a warlike state between Indigenous and settler presences.

Much is known of this early period due to the oral histories of Yanyuwa themselves and also the documentations of ethnographers and missionaries who passed through the region during the 1900s. Yanyuwa elders recall stories told to them by the old people of first encounters with white people, including stories of shootings and the harassment of Aboriginal women. With the arrival of Europeans there also came an altogether different economy and pattern of land and sea use. In the 1860s, there were a variety of colonial interests pushing for the annexation and settlement of the north, where there was seen to be opportunity for an expanding pastoral industry (Anderson and Monteath 2008: 2). By 1911, mining exploration had begun in the region, along the McArthur River, and by the 1920s Europeans had taken over the local trepang industry (as established by Macassan fishermen who came from Sulawesi). By 1931, land was leased for a salt works industry on the eastern fringe of Yanyuwa country.

A most significant and lasting change for Yanyuwa was the move from bush life to town life, which brought a break in the practice of living on and regularly communing with country. This occurred from the 1930s and reached its peak in the 1950s (Baker 1999: 5) as pastoral leases and stations heavily partitioned and alienated large portions of land throughout the region. The move away from communal camp life began with infrequent visits to the township of Borroloola, and resulted in increasingly centralised and permanent settlement by the 1960s. In terms of residing more permanently in the Borroloola area, Yanyuwa established a large camp at the place of Malarndarri, across the river from the current township of Borroloola. From 1968, a series of events impacted significantly upon the community and community life, ultimately affecting living patterns, knowledge-sharing and language use. A virulent form of influenza spread through the camp causing the death of several elders (Bradley et al. 1992: 27; Bradley with Yanyuwa Families 2016). The epidemic was understood as a consequence of cattle disturbing a sickness site associated with areas contained by the Bing Bong pastoral lease. In the aftermath of this widespread sickness and death, the camp was abandoned and several scattered camps were established, until further deaths prompted additional disbandment and a permanent move into the township.

A further breakdown in camp life and time spent living on country was accelerated once again in 1974, when major floods led to further residential dislocation and growing dependency on the offerings of town life. For Yanyuwa, this and the influenza were deemed communicative events, which spoke to the consequences of harmful agents disordering country. Disordering events such as site disturbance, aggressive settler interactions with place and its elemental parts, and disregard for the Law that constitutes country trigger responses from ancestral and multifarious agents. These responses may take the form of floods, extreme droughts, non-human species death and decline, and also human death and decline. The impact of dislocation from country and places of importance was residential fracturing, increased separation from and anxiety about country and a subsequent breakdown in Yanyuwa language proficiency (with increased use of Kriol in the camps at the expense of the Yanyuwa language). Ceremonial life and other cultural practices associated with being on country, including hunting and gathering techniques and material cultures, songlines and orations were also compromised (Bradley et al. 1992: 26–27; Bradley with Yanyuwa Families 2016). In addition to residential shifts, colonial education models and content, and impositions of European 'social life' have had a role to play in redefining people's relationships to place and country.

For Yanyuwa, both young and old, Bing Bong has always been valued as an important place in terms of continuity with the past (Bradley 1997: 84). This continuity was fundamentally challenged by the granting of the Bing Bong Pastoral Lease and further threatened when the federal government authorised the sale of Bing Bong Pastoral Station to Mount Isa Mine (MIM) in 1976. This took place, despite Yanyuwa expressions of interest in acquiring the station so as to safeguard their homelands with greater certainty (Bradley 1997: 85). Mount Isa Mine would then consolidate three large pastoral leases, Bing Bong, McArthur River and Tawallah leases, generating a land holding of massive proportions. The sale of Bing Bong Station to a mining interest significantly affected the wider Yanyuwa community, as it signalled further alienation from country, along with the threat of place destruction, elemental decay and toxicity, understood to be the greater effects of mining industries. Yanyuwa elder Nero Timothy captured this concern and worry when he stated:

> Yanyuwa people really belong to Bing Bong and we care about what we heard on the news that the mining company got that place so people they really worry about it and they still camping out there. Yanyuwa people meet here and have big dance and even go and teach our children how to hunt, show them how to get turtle, dugong, and kangaroo. So you see that really bad and we want to try and get away from town too because this town will be a mining centre and we know this town will bring a real lot of bad people with different stuff in here and then they'll ruin this place. Ruin our people's lives and we can't have any freedom, any freedom at all, so you see. And this is what we want you to see, the point about this place Bing Bong. So I don't know why the miners got hold of it, I don't know for what reason.
>
> (Nero Timothy 1978, in Bradley 1997: 85–87)

By 1976, MIM, in holding three pastoral leases over portions of Yanyuwa country, had succeeded in its aim to alienate Indigenous landowners from country and begin a widespread project of mining extraction and raising cattle as part of its ongoing investment in the region's mineral and pastoral industries (Kearney 2009b: 200–203). By holding pastoral leases over this land, MIM created a stronghold that could be used as a bargaining tool in the event of Aboriginal claims to land and sea country (Bradley 1997: 85). This became an issue in 1977 with the Borroloola Land Claim, the first claim for land restitution to be lodged in the Northern Territory under the *Aboriginal Land Rights Act (Northern Territory) 1976* (see Avery and McLaughlin 1977; also Aboriginal Land Commissioner 1979). Able to passionately articulate the loss and destructions taking place across country, Yanyuwa have never accepted their forced removal or alienation from these places. Nor were they or have they ever conceded to the inevitability of the rapidly moving destructions that came with the pastoral and mining industries in this region. By seeking to contest this via legal recourse, the 1977 land claim was followed by further claims under the Act in 1992 and 2000, and a Native Title claim under federal law in 2015 (see Bradley et al. 1992; Bardon 2015a).[4] These claims pertain to lands and waters that cover vast areas of mainland, coastal and island country.

The evidentiary burden on Yanyuwa claimants has been huge and has led to an impressive documentation of ancestral presence and knowledge of place and country more broadly as part of the evidentiary requirements of the Land Rights Act. In the fullest sense, Yanyuwa have enacted and evidenced kinship with place, for nearly four decades, to an audience of non-Indigenous lawyers, anthropologists and government bodies. In doing so, they have not faltered in the assertion that this is the enactment of their moral obligation to the old people and is a vital part of their very existence. By way of this land claim history, Yanyuwa have won back recognition of their exclusive and freehold Indigenous title to parts of country, both the offshore islands and smaller sections of clan country on the mainland. Today this is held by Yanyuwa families as inalienable title, to the exclusion of all others (Young 2015). That said, the lands and waters consolidated under both pastoral and mining leases remain off limits, and subject to unmitigated harm.

The area around Bing Bong became heavily implicated in the 1977 Borroloola Land Claim in part due to the proposed plans for mining expansion by MIM. As background noise to the claim, MIM had developed proposals for a railway from the mine site to a proposed township and port on Centre Island, in the Gulf of Carpentaria. Access for the railway was to pass through Bing Bong, thus implicating this place in the political frenzy that saw the Northern Territory government and MIM actively oppose the Yanyuwa claim. Evidence from Indigenous claimants was heard before a largely hostile audience of non-Indigenous persons, government and mining representatives in Borroloola (Young 2015: 2). There was little effective use of interpreters, women were not permitted to give evidence and there were no site visits to the areas under claim (Young 2015: 2). While the Aboriginal Land Commissioner, Justice Toohey, recommended the granting of Aboriginal freehold title over land around Borroloola to West Island and Vanderlin Island, the recognition of Aboriginal land title did not extend to other islands throughout the region, including Southwest Island, Centre Island and North Island (Young 2015: 2). The stinging part of this 1978 decision was, that for Yanyuwa, "these islands were the core of the claim", and were also targeted areas for the Mount Isa Mine's planned construction of a railway, town and port facility (Young 2015: 2).

What ensued were counterclaims in 1979, and 1992, complicated by a series of illegal acts by the Northern Territory government to dispute and contest the legitimacy of Yanyuwa claims. The progression of mining interests above and beyond those of the region's Indigenous owners also remained a strong theme throughout this time. Then in 1992, Mount Isa Mines announced it was to embark on large-scale mining operations at McArthur River mine, the primary mine site located inland from Borroloola on Gudanji country. In 1993, the Northern Land Council, acting on behalf of the Yanyuwa and Gudanji traditional owners, "wrote to the Northern Territory and Commonwealth governments seeking to be heard on aspects of the proposal, particularly social impacts on the Aboriginal people at Borroloola and environmental impacts on the McArthur River. The letter was ignored" (Young 2015: 2).

Government inducements were forthcoming and a campaign of coercion began. Yanyuwa, who held ongoing concerns as to the environmental and cultural impacts of the mine and the flow-on effects of river contamination into their saltwater country, faced the realisation that as the Commonwealth and Northern Territory governments ratified the mining approvals with special legislation, the mine would go ahead. Bing Bong in turn became the primary site of interest for the mine's mainland port and loading facility. Bing Bong was destined to house an enclosed concentrate storage facility, barge-loading wharf and bulk carrier, accommodations and sites for haulage equipment. This meant dredging an offshore channel through which the bulk carrier could transport concentrate to an offshore transfer zone, where upon meeting buyer ships, the lead and zinc concentrate would be funnelled through a loading chute several meters above the sea (Glencore 2015). It was a fait accompli. The impact of this decision and the inevitable construction of the loading facility in 1994 forms the substance of the proceeding discussion of place harms.

Despite years of negotiation, and Yanyuwa attempts to constrain the incoming destructive forces of mining companies and governments intent on profiting from these activities, the people and places of the southwest Gulf of Carpentaria have been severely harmed. Events have followed a predictable pattern of place harm, whereby "they have cut the ground completely and deeply; they have broken all the trees and wiped out completely the camps of the old people ... they are totally gone" (Dinah Norman Marrngawi with Roddy Harvey Bayuma-Birribalanja in Bradley 1997: 95). The possibility of the country being damaged as a result of mining activities remains a huge concern for people, particularly now, as they have seen the direct impact that mining and its associated production phases have had on country as a sentient co-presence. For many, these activities contravene every element of Yanyuwa Law, and the violent agents of mining's operations damage country completely, they leave nothing and those who intently lead this harm provide no answers (Dinah Norman Marrngawi in Kearney 2009b: 206). For Amy Friday Bajamalanya, places once for Yanyuwa kin are now only accessible through the permissions of "whitefellas, where you gotta ask them for permission [to visit]". For elder Pyro Dirdiyalma, there is so much that is "properly lost now" (Kearney unpublished fieldnotes, 2002).

The giant who rips the guts out of Bing Bong

Before moving on, it must be said that I continue this discussion on place harm around Bing Bong with some degree of trepidation. The reason for this is that Yanyuwa, and their neighbours, Gudanji and Garrwa, have a complicated relationship with the mining company responsible for the destruction of parts of their respective homelands. With specific reference to Yanyuwa experience, today many work at the mine and thus source their living from Glencore (formerly known as Xstrata, the company that purchased Mount Isa Mines in 2003). The company that owns and operates the mine is woven tightly into the fabric of everyday life for the Indigenous community of Borroloola in ways that are complicated and defy a simple determination that, (1) people do not want the mine to exist, and (2) there is a simple solution to be found in the cessation of mining activity within the region. Sentiments concerning the mine vary across generations, and reflect individual experiences. That said, from widespread discussions and long-term ethnography, I have yet to record an Indigenous perspective that endorses the mining company's destruction of place, nor acquiesces to the environmental and ecological destruction that has already occurred.

Not dissimilar to the tensions experienced by Indigenous communities elsewhere in the world (see Kirsch 2007 for a discussion of the Ok Tedi Mine in Papua New Guinea), people are often forced to choose between degradation of the ancestral place world, and economic opportunities and vital community services in locations where deep poverty and disadvantage is a lingering consequence of settler colonialism (see Kirsch 2007: 314). Not wanting to dispossess people of agency and their right to choose the nature of economic investments and participation that may bring about greater life opportunities, it is important to contextualise people's views of and relationships to the mine relative to the opportunities available, the disadvantage that is palpable upon entering the township of Borroloola and the overall failure of the Australian government to remedy these situations. It is Glencore that provide the residents of Borroloola with some of the region's only employment opportunities. The company has also funded a much-needed local swimming pool and dialysis unit.[5]

Glencore also assist the community by providing the resources for air travel from this remote part of Australia in the event that Indigenous family members must travel to visit sick relatives or attend funerals. Local funerals for Indigenous community members in Borroloola are also supported by funds from Glencore. Each of these provisions inadvertently alleviates the Australian government of its responsibility to provide community services and health care as the provisions for a good life to all of its citizens, particularly those living in remote Indigenous communities. The dialysis unit that now operates in Borroloola provides a service that is vital for the survival of many Indigenous residents, who succumb to kidney disease and kidney failure at an unprecedented and shocking rate.

Early death from kidney disease has become one of the highest causes of death among Indigenous Australians and strikes hard at persons of all ages (see Australian Indigenous Health*InfoNet* 2015; Spencer et al. 1998). The requirements of dialysis treatment have meant that many Yanyuwa must leave their families to live in the city of Darwin, some 1,100 kilometres away, where they can receive regular and life-saving treatment. They do this in isolation and in often rather lonely situations. The Borroloola dialysis unit was requested as a provision that would allow people to stay close to family and country while receiving treatment; as both are vital to a healthy Yanyuwa existence. In 2009 Glencore (at the time Xstrata plc) agreed to fund a dialysis unit (McArthur River Mining 2009: 5). Thus, for some Indigenous residents in Borroloola, their very survival is dependent on the mine. The cruel reality of this in a settler colony cannot be overlooked or understated. It plunges deep into the darkness of cultural wounding, and returns this discussion to an awareness of the nested ecology in which one sickness can lead to another. The violence of everyday life that manifests as poor health and shocking mortality rates in this and other remote communities mirrors the violence of what is being done to country as a sentient co-presence.

In sum, there is complexity in how mining companies and their operations are negotiated by Indigenous groups (see Kirsch 2007: 314) and local decision-making is rendered part of a deeply wounded scape supported by settler colonialism and neoliberalism. I am not attempting to simplify this, rather to contextualise it. As Gloria Friday (pers. comm., 3 October 2015), an Indigenous woman and long-term resident of Borroloola who is married to Yanyuwa elder Graham Friday, explains, "it's complicated, because us people here in Borroloola need that mine, that director, he's ok, he helps us out when family is sick, when we have to go see them. The mine gives us jobs, and we can't be cheeky about that. Some people, those young ones, they get upset, but what are we going to do?" Then in a moment's breath, a family member sitting alongside us, responds to Gloria's comments, saying, "it's just all shame, big shame you know, that place Bing Bong used to be all for our families, for camping and the old people, full of Law, but bit by bit they been kill 'im". Few would understand the dilemma this community faces.

The Bing Bong experience

The Bing Bong loading facility and actual mine site are bound by the experience of disorder. Although separated by a distance of approximately 120 kilometres, they are physically linked by a bitumen road and relationally bound through ontological awareness for the residents of Borroloola. Semi-trailer 'road trains' transport the concentrate from the mine site to the coastal port regularly. On a typical day, a single driver may make three return trips, and four trucks will be operating at all times (McArthur River Mining 2015: 8). Each vehicle can be loaded with up to 120 tonnes of zinc concentrate (McArthur River Mining 2015: 8). These vehicles cross over country that is thoroughly mapped with ancestral activity, which is narrated through oral testimony as the country of the old people and the country of present generations who have camped, walked, hunted, danced and sung the length of their homelands.

According to Yanyuwa epistemology, the places that run the length of country commune with one another, and are known to stand in kinship relations with one another; for example, two places may be brothers, or a sequence of place may come together to form the body parts of one larger ancestral being. These places become kin and are relationally bound to one another through the journeying of ancestral beings. It may be that the ancestor stopped in one place before moving on to create another, their breath may be left in place, their weapons, their song. Complexity of interactions and endless relational encounters is what makes up the very substance of the place world. Such encounters of movement (journeying) and stasis (sitting down as place) are what define the relationships that bind powerful places on Yanyuwa country. Wulibirra and Muluwa are one example of this. These two brothers sing out to one another as part of the ancestral journey, which led to their creation. When one looks west from Muluwa, across ten kilometres of sea there are, rearing up from the sea, the cliffs of Red Bluff, Wulibirra. In Yanyuwa, Muluwa is the older brother and Wulibirra the younger brother. Collectively these two places are called by the kin term *nyinkarra* – younger and older brother (see Bradley and Kearney 2011: 37).

Places are wrapped in such relational meaning. They communicate back and forth to one another, with the events taking place on one part of country having an effect on the other, as a nested ecology. The relational absorption of harm, however, goes even further, to include the non-human animals that make up part of the nested ecology of place. No place, person or element of life exists in isolation, nor is any part of country empty and without kinship. To speak of the violence that has occurred at Makukula, Wimanda, Arrinyanda, Wurrulwiji and Mawuli is to directly implicate actions that occur elsewhere, namely at the mine site and at multiple places in between and a vast number of elemental presences that may be impacted.

In 1995, the Bing Bong loading facility placed its first shipment of concentrate onto the *Aburri*,[6] a large ship that departs the loading facility to meet the buyer ship 20 nautical miles offshore (McArthur River Mine 2015: 16) (see Figures 2.9 and 2.10). In 2002, an application was made to expand operations

Figure 2.9 Bing Bong Port Facility, Glencore McArthur River Mine, 2015 (Photo source: Amanda Kearney).

Figure 2.10 Bing Bong Port Facility, Glencore McArthur River Mine, 2015 (Photo source: Amanda Kearney).

at the mine, from underground production to open-cut mining (Young 2015: 3). Tension between the company and Indigenous community members was high throughout the period of proposed expansion, fuelled by grave community concerns regarding the destruction of sacred sites, dangers associated with diverting the major river system, threats of poisoning in the estuarine and marine environment and contamination of waterways with heavy metals. The potential overturning of road trains on the way to Bing Bong, carrying large quantities of concentrate, as well as the destruction of the coastal areas and mangroves also weighed heavily on Yanyuwa minds (see Bradley and Yanyuwa Families 2007: 47, 63). Adding to the tension were ongoing health concerns among those locals working on the Aburri, as well as locals and visitors who fish throughout the area. To date there remains much conjecture as to the likelihood of exposure to high levels of lead and zinc concentrate. Reports of community concerns describe the local Indigenous perspective on harms thus far done, with people noticing "sickness in dugongs and turtles downstream of the mine soon after it began operating" (Polidor and Tindall 2006). "Community members reported seeing thousands of dead fish in the bay and shellfish dead on the sand ... more erosion; [and] the disappearance of fish, turtles, coral reef and sea grass" (Polidor and Tindall 2006).

In a protracted sequence of events, the move to fully operational open-cut mining eventuated in 2009. With much back and forth as to the legitimacy of approvals for the expansion, the case entered the Northern Territory Supreme Court and later the Australian High Court. Environmental impact statements were scrutinised and Indigenous owners demonstrated, presenting their concerns to parliament and to a wider Australian audience across social and news media platforms. In order to expand operations, the mining company sought to divert the McArthur River, the lifeblood of this region, a powerful Rainbow Serpent ancestral being (Howell 2007, 2008). The effects on maritime heritage, marine biodiversity, bird populations, endangered fish and fish breeding grounds were all cited as cause for concern and prevailing justification to surely deny the request to expand.

Alongside this ran a passionate plea to see country as more than a series of extractive opportunities; a call to witness these places as the old people, as ancestors, as kin and as profoundly important co-presences that support the overall health and well-being of an entire community. Gudanji, Garrwa and Yanyuwa alike called on the wider community to think of and imagine the consequences of cutting the Rainbow Serpent's body, of digging so deeply that the old people and the region's non-human species might die (see McArthur River Mining 2007 for the text of a speech given by Yanyuwa woman Malarndirri McCarthy [formerly Barbara McCarthy]). That this failed to illicit shock and response to the point of restricting the mine's operations reveals an axiological crisis and subsequent retreat among those responsible for decision-making. There were significant risks involved in this expansion, none greater than the absolute destruction of country as a sentient presence. Despite this, it was approved.

The biography of place harm continues as Bing Bong is still today called upon to support this extractive industry. It has paid a huge price for doing so. Across Makukula, Wimanda, Arrinyanda, Wurrulwiji and Mawuli, people report significant changes to country. They report poisonings of non-human species in the region, they worry that dust from the ore body is carried from the mine site to the port. They fear for the spillage of concentrate into the sea as buyer ships receive the ore. People ask, will this go into the sea? Will it kill the seagrass?[7] Will the dust from the ore body enter the ecosystem and be ingested by the sea turtles? (Bradley and Yanyuwa Families 2007: 47). Yanyuwa fear that turtles are killed and injured when the Aburri barge manoeuvres itself at the loading facility, hit by the thrusters and propellers of the barge as it docks and that hatchlings are detrimentally affected by the bright lights that illuminate the loading facility (Bradley and Yanyuwa Families 2007: 47, 63). There have been huge changes to the physical form of place through dredging, and changes in tidal patterns. Yanyuwa also express concern regarding a prevailing threat that annual flooding or cyclone activity may one day overcome the storage facilities where the bulk concentrate is stored, thus leaching this toxin all over country (Bradley and Yanyuwa Families 2007: 63).

The spread of harm has grown as the company has expanded production and adopted a 'remediate later' approach to its operations (see Bardon 2015b, 2015c). As the company took a share fall in 2015, major concern has been raised as to environmental liability and the long-term effects of the mine and port facility (ABC News 2015a). The region's residents have, in the past year, raised further health concerns and concerns over county due to reactive waste rock from the mine and leaking tailings dams. The Northern Territory planning minister David Tollner and chief minister Adam Giles threatened to close the mine if environmental practices were not improved. They also requested the company's environmental (remediation) bond be increased, stating: "We have been adamant that unless Glencore fixes its environmental procedures and practices we will close the mine" (in Pearson 2015). "We will not stand for an environmental bond that does not support rehabilitation at the end of the mine life. We will not support procedures that put potentially high risk to the environment" (Pearson 2015; see also Bardon 2015c).

All mining companies in the Northern Territory are required to lodge a bond with the government that would cover 100 per cent of their site's final remediation cost at every stage of the mine's life. Negotiations on this matter continued throughout 2015 in the wake of a huge waste rock pile that began smouldering in December 2013 on the mine site. For several months throughout 2014, the reactive iron sulphide gushed large smoke plumes across the region (see Bardon 2014). Gudanji, Garrwa and Yanyuwa community members demanded that Glencore stop the smoke plume before it posed a serious threat to the entire region. For months, Glencore said it was trying to extinguish the fire but could not say when it would be put out, acknowledging that the "rock pile began spontaneously combusting when pyrite iron sulphide was placed into its top layers". The company attempted to manage the problem by coating the rocks with lime and clay but this did not stop combustion (see Bardon 2014).

On these matters, Yanyuwa man David Harvey reported his concern, stating: "Our generation and our grandfathers been fighting for this country to keep it together, now they've come and destroyed this country" (in Bardon 2014). The smoke plumes from the combusting rock pile caused further concern when the Department of Primary Industries announced that cattle had been potentially contaminated with dangerous toxins after wandering onto the mine site (ABC News 2015b). The company was instructed that some 400 cattle in a 100 square kilometre radius of the mine be quarantined or killed. Half the cattle were shot and the remainder quarantined amid concern about the contamination of other animals in the remote region (ABC News 2015b).

The battle over remediation and the combusting waste rock pile has shadowed events simultaneously taking place at Bing Bong. Western Desert Resources (WDR) began operations in the Gulf region in 2014. In 2011, WDR signed a memorandum of understanding with Xstrata (Glencore) to test the feasibility of expanding the Bing Bong loading facility to accommodate the transfer iron ore from its neighbouring iron ore project at Roper Bar in the Northern Territory (Latimer 2011). This would require the construction of a private, sealed haul road from the mine site to the Bing Bong Facility, approximately 165 kilometres in length. Despite company claims that there is little in the way of sacred sites or Indigenous interests in the area transected by the proposed haul road, Yanyuwa are adamant that this is not the case. Ancestral beings move fully and actively throughout this country, thus risk being harmed by WDR's actions. Haul trucks would then travel to the Bing Bong location where the ore deposit would be stored ahead of its being conveyed to the Bing Bong barge loader via an open-air conveyer belt. The haul road was built and transfers began in early 2014. By September 2014, the company had entered into voluntary administration (Brown and McCarthy 2014). During the ten months that it operated, the company was plagued by controversy and critiqued for not adhering to world standards in environmental protection.

WDR's facilities and operations at Bing Bong represent a secondary attack on place. For Graham Friday, Yanyuwa elder and head of the li-Anthawirriyarra Indigenous Sea Ranger Program, "that Western Desert mine has gone and buggered up that place, properly buggered it up, now it's really ruined" (Graham Friday pers. comm., 3 October 2015). In October 2014, the company, by then in receivership, was issued a direction by the Northern Territory Environmental Protection Authority, under Section 72 of the *Waste Management and Pollution Control Act* when environmental concerns were identified at the mine site, along the haul road leading to the Bing Bong loading facility and at the Bing Bong loading facility itself (Northern Territory Environmental Protection Authority 2014). They were instructed to "take all reasonable measure to divert storm water away from and prevent flooding of mining activities; install and rectify appropriate sedimentation ponds and associated drainage systems to service mining activity areas; and take all reasonable measures to prevent erosion and sediment runoff at bridge approaches along the haul road" (Northern Territory Environmental Protection Authority 2014). Furthermore, specific instructions were given to "investigate and implement appropriate measures (that should not cause additional environmental impact) to clean-up the deposition of iron ore material on the tidal flats" (Northern Territory Environmental Protection Authority 2014). Western Desert Resources has been no stranger to controversy over its environmental management practices, with traditional owners and fishermen raising concerns over iron ore dust spilling from the company's loading facilities at the Bing Bong port (see Brown 2014; Brown and McCarthy 2014).

That these mining companies have been able to act with such disregard for country and the Indigenous lives it supports is shocking. The principles of remediation, of monetary bonds and instructions after the fact are weak attempts to undo what has already taken place. The destructions that have been enacted with intent at Bing Bong, cannot be easily remedied, as time has shown. A number of scientific assessments have been carried out around the site of the loading facility, both offshore and in the immediate coastal environment, in an effort to examine the likelihood and/or extent of harm caused to the local environments in which mining operations have occurred and continue to occur (see, for example, Munksgaard and Parry 2001; Munksgaard et al. 2003). Some have tested for lead and zinc concentrates, trace metals, effects of heavy metals on seagrass beds, demographic and health parameters of green sea turtles in the Gulf of Carpentaria. Glencore has funded some of these studies, while others have been independently carried out. Very few have been undertaken in collaboration with Indigenous traditional owners of the lands and waters under investigation (with the noted exception of a study by Hamann et al. 2006, which is co-authored with two Yanyuwa community members).

In the context of this book, it is suffice to acknowledge the existence of these scientific studies and to summarise that while some find elevated levels of contaminants and cause for concern in the local ecology, others do not. That said, those findings, which refute the toxicity argument, do not function to undermine Yanyuwa, Gudanji and Garrwa claims of the violent and wounding effects of the mine and operations at the Bing Bong port facility. This book is not concerned with testing the 'legitimacy' of Indigenous claims to place harm and the sickness the mining industry has instated. To do so would be an act of epistemological violence in which a singular vision of knowledge, evidence and truth is promoted; namely a non-Indigenous vision over that of Indigenous knowledge. Yanyuwa configure their concerns for country and the places of the old people in ways that are largely untestable using scientific methods and Western rationalism, for they reside deep within the psyche, in fact deep within the bones, the least corruptible part of the human body, as is the case with an individual's place-derived spirit child.

The wounding that has occurred in place, and by default within people, expresses a breach of Yanyuwa Law. It is this Law and its adherence or disregard, that informs the emotional geography of care and concern, and the experiences of Yanyuwa country within a nested ecology. If pressed to respond to the questions, 'how do you test the health of country?' and 'how do you know when country is sick?', the response is that Yanyuwa are best positioned to determine one way or the other, due to the very nature of their kinship with place and capacity to commune with multifarious agents in and of country. Furthermore, one need only look at the disordering and destruction that is evident at Bing Bong, and the hardships being endured by Yanyuwa as a result of this to gain a sense that something is deeply awry.

The signs of place harm and deep wounding are evident for those who know how to read country and who know the inherent order that characterises these places. There is evidence of ecocide, seen in the absence of kangaroos, shells, turtle and dugong remains (washed up on the beaches of Wurrulwiji, and Arrinyanda as a result of storm activity – yet indicative of healthy populations out at sea). These were once the tell-tale signs that place was healthy, functioning as it should. The lack of such creatures and presences is traced to the ever-expanding presence and influence of the port and loading facility. The decline and disappearance of non-human animal kin is a key indicator for Yanyuwa that disorder has descended on place, caused by breaches of Law. The greatest breach of Law comes from its denial and the failure to recognise kinship with place. There is a deep axiological schism in the ethos of the mining operations and that of Yanyuwa Law, and their differences cannot be overstated. The crisis of neoliberalism, modernity and coloniality is such that kinship is fundamentally denied both through praxis and ontology, having already found its foundations epistemologically. Not only do these operations deny any inherent human kinship with place, they also obliterate the right for any other creatures to have kinship with place.

Ecocide marks the axiological crisis of settler colonialism. Throughout the history of colonial economies in the Gulf of Carpentaria there have been periods of dramatic species decline, disappearances, reported mutation and hybridity (Seton and Bradley 2004; Trigger 2008; see also Rose 2002 for a discussion of similar experiences elsewhere in the Northern Territory). These non-human species are kin and are a vital part of the elemental aspects of place. They are nested within the ecology as manifestations and embodiments of the old people, resources and food. Species ranging from blue-tongue lizards, to goannas, quolls and crabs have all at some point been deemed, by Yanyuwa, as 'gone', 'finished' or 'hidden' as a result of place harm. The agents held responsible for these captivities and disappearances are the full range of invading presences that fail to recognise Yanyuwa ancestral Law in place, as well as existing kinship and morality in place.

These agents of harm have been early settlers, tourists, mining companies and feral species, including cane toads (introduced to Australia by the Commonwealth Scientific and Industrial Research Organisation), cats, pigs and goats. They have ushered in elemental decay and ecological decline through a disregard of place value and Law. So too have they denied the kinship that renders relationships between human and non-human species as meaningful in the everyday and spiritual lives of Indigenous people in this region. The desire to culturally wound Indigenous owners of lands and waters throughout Australia has been so long-running and is so deeply ingrained in the nation's psyche that the effects of harm-doing have extended to include the entire epistemological, ontological and axiological realms of Indigenous people's lives. The killing of non-human kin, the destruction of place, annihilation of old people and ancestral beings, are all done in the service of an alleged march towards 'progress' by way of 'regional development'. That this comes at a cost to Yanyuwa, their place world and its Law is overlooked when the axiological crisis omits the possibility of place as a sentient co-presence.

Reflecting on destruction, disorder, toxicity and elemental decay

Pastoral and mining practices have destroyed, disordered and toxified place. Around Bing Bong, this has been experienced through a series of events, beginning with the killing of Wilangarra people, then the alienation of Yanyuwa from their homelands. Next came the granting of a pastoral lease, and the construction of a large-scale port and loading facility to support mining operations. These heavily undermined a Yanyuwa presence in the region, and have threatened the authority of Yanyuwa Law to govern the nature of human engagements with place. In the very early days of the Bing Bong port development, Yanyuwa elder, Mussolini Harvey, spoke regretfully of the disturbance this caused to the old people and the spirit children (*ardirri*) who reside there:

> Lots of people saw them, especially at night. They were many small spirit children darting across the road, coming from both sides, then the following night four old spirits were seen bowed over, hands behind their backs and dressed in bushel bags [hessian sackcloth]. The spirits were looking for new homes, they can hide under logs, make themselves small, they will find a new home ... but makes me think ... how many of the poor buggers died from all that work ... makes me really, really sorry.
>
> (Mussolini Harvey, pers. comm. Bradley 1994
[unpublished fieldnotes], cited in Kearney 2010: 113)

Upon visiting Warangala, a place near Bing Bong, in the midst of earthworks and port construction, Roddy Harvey Bayuma-Birribalanja, who sadly passed away in 2015, and whose *ardirri* (spirit child) originates from Wurrulwiji, exclaimed, with a sense of deep physical and emotional pain:

> Ah dear me! This is all too much I am telling you! They have cut you! They have emptied you! My stomach burns with shame for you! I walked here when I was young with my father, with my mother, my father found me in this country, I come from this country, my spiritual essence is from here, this country is the point of my origin. I am crying, tears are falling, they have cut you deeply, so far down, they have flattened you, wiped you out completely, I am finished now, I just had to tell you this!
>
> (Bradley 1997: 95)

As Rose (2008: 111) writes: "Country is a living entity with a yesterday, today and tomorrow, with a consciousness, and a will toward life." "Persons are located in country, and country too can be conceptualized as a person in the sense that it is self-organizing, sentient, autonomous, and interdependent with other countries" (Rose 2008: 110). Roddy's exclamation, as she stood looking at the suffering of her place kin, reaches out to a long line of ancestors. It contains a deep sadness, a grief at what has been done to this place and a frustration that ecological time and sequential time are in battle with one another in the context of place harm. The damages done can rarely be remedied in time to keep country strong and up high. As her spirit child comes from here, and resides deep within her bones, the pain she felt must have been palpable. This brings forth the realisation that perhaps some harms cannot be undone.

Roddy's recent death has brought a great sadness to Yanyuwa families and all those who knew her, in a way that the destruction of Bing Bong brought her a great sadness during her final months. In April of 2015, when I last saw this old lady (who I refer to as *ngabuji* – father's mother), she remarked to me that "Bing Bong is proper dead now, that place, they finally been kill 'im" (pers. comm. Roddy Harvey Bayuma-Birribalanja, 5 April 2015). I had never heard a Yanyuwa person so completely decisively speak of place death. I have been reluctant to write of this, not only because of the depth of sadness it induces, but because of the considerable political weight it carries, certainly within the Yanyuwa community itself, where a declaration of this kind must be carefully made, relative to a wider community of persons who stand in a clan-based relationship to place. But also I am mindful of how such statements can be misused by a political system that is built on the principles of settler colonialism and that has propagated a language of deficit concerning Indigenous communities and their cultural lives (Kearney 2014: 78).

Roddy's comments do not equate place death with the disappearance of Indigenous pedagogies of place, nor the cessation of Indigenous ownership of these areas as parts of ancestral territory. Her comments instead reinforce the extent to which these matter and prevail, despite the failure of outside agents to recognise them. What her comments also raise is the question of whether place can return from the great harms of cultural wounding. It is a question often asked of the Borroloola community more broadly, but in the context of this book is located within a discussion of the place world (Kearney 2014). What might be a death point for place? And can place survive the greater effects of cultural wounding that manifest in the form of intent-laden destruction, disorder and toxicity?

The effects of cultural wounding on Bing Bong convey the absolute limits of disregard for Indigenous Australians and their ancestral homelands. The agents of harm have been defined by the extent of their moral disengagement, not simply as a register of their willingness or unwillingness to do 'right or wrong', 'good or bad' but by convincing themselves that the ethical standards called for by Yanyuwa Law and other Indigenous laws that exist in place do not apply to them. By separating themselves out from the nested ecology of country, which has its own order and thus its own set of ethical principles concerning all life and its elements (inclusive of the place world), agents of harm act with disrepute. Whether declared as moral disengagement or 'rational ignorance', extractive industries do great harm to Indigenous homelands across the globe. Refraining from acquiring knowledge, in the form of kinship and a relational encounter with place, because the cost of education might be an inability to continue with acts of place harm is a distinctive marker of much violence. This is most certainly the case for Bing Bong and the recent harmful acts, which have caused its possible death.

That agents of harm are permitted to act with moral disregard and rational ignorance is no accident. Their actions are embroiled in an ongoing tension that sees the dismissal of Indigenous rights and authority. This is an authority led by epistemological, ontological and axiological systems of distinction and substance, which require full accountability and relational awareness. When all is kin, and when human life is nested into the lives of all others, even that of place, how are violence, destruction and killing possible? (Rose 2013: 141). With axiological crises, as described in this chapter, the problem lies in value nihilism and not knowing or caring about 'how to be' in place, "what kind of life is worth living and preserving". When the unrestrained disregard of people and other multifarious agents in the world is allowed to proliferate, what conflicts and survival crises are we willing to accept? Remembering the role of the witness, I return to Rose, who challenges the crisis.

> This brings us to the matter of witness, which includes and exceeds our capacity to attend to silences that are themselves stories, and to stories that are themselves philosophies. In an ethics of mutuality, we would strive to keep life flourishing not only by defending the lives of all of our kin, but by taking a stand for the meaningfulness of both life and death, and for a world where nothing belongs on the rubbish dump.
>
> (Rose 2013: 146)

The following Chapters 3, 4 and 5 usher this discussion into a third phase of reflection. They operate as a diagnostic pathway that, when followed, reveals key aspects of the methods and motivations that bring about violence in place and associations with cultural wounding. The wounding of place is framed in three parts: first as the destruction and designification of place as a physical locale; second, wounding as a culturally constructed vision reflecting where people and cultures are held to belong; and third, through the killing and decline of ecosystems and non-human species. By focusing on the suffering of place and non-human species in the midst of cultural conflict and violence, as measured by death, disappearance, extinction and decline, the relational experience that is cultural wounding is rendered explicit for the reader.

Notes

1 *Jungkayi* are the children of the women of the patriline who are associated with the Dreamings and ceremony of particular parts of country (Bradley et al. 1992: 159–160; Bradley with Yanyuwa Families 2016). One inherits the relationship of *jungkayi* through their *kudjaka* (mother), *mimi* (mother's father), *ngabuji* (father's mother and father's mother's brother).

2 Bauman (2002: 213) has written an excellent account of shame as a very particular emotion within the context of Indigenous Australian community life. She explains that "[s]hame is the most powerful emotion in Aboriginal life and is the province of the socialised being (Stanner 1979: 95). It is simultaneously a force of social control and an individual emotional improvisation: whilst preying on emotion, it is paradoxically reproduced by emotion." Shame invokes fear, the "fear of shame" (Hamilton 1979: 336, cited in Bauman 2002: 213), and in the context of this place harm discussion, this fear of shame comes from the possibility that place harm reveals a breach of law, or failure to enact kinship and responsibilities to place. As such, it acts as "[a] regulatory force of negative evaluation and a mechanism of social control" (Bauman 2002: 213; see also Burbank 1985; Hiatt 1978; Morris 1989).

3 Yanyuwa Law organises all people and all elements of life into a series of clan groupings. There are four clan groups including, Wuyaliya, Wurdaliya, Mambaliya and Rrumburriya.

4 While the 2000 claim was heard and a decision reached, in favour of recognising Yanyuwa land tenure, the Northern Territory government never honoured the finding of Aboriginal land title by the Aboriginal Land Commissioner.

5 The importance of a local pool should not be understated and goes beyond providing much-needed relief from the intensity of a north Australian and monsoonal summer. It provides young people with one of the few exciting distractions in town. Pools are also regarded by medical professionals as an important addition to remote Indigenous community settings. To quote a major study on the impact of swimming pools on Indigenous children's health in remote communities, in particular ear and skin problems: "Indigenous Australian children have very high rates of pyoderma (pus-producing skin lesions) and otitis media (glue ear). In some communities as many as 70 per cent of children have been found to have skin sores, at any one time" (Lehmann et al. 2003: 415). The findings of this study have shown that: "Swimming pools in remote communities have been associated with reduced prevalence of perforations of the tympanic membranes and skin sores, which could result in long-term benefits through reduction in chronic disease burden and improved educational and social outcomes" (Lehmann et al. 2003: 415–419).

6 The ship was given the name *Aburri*. This is a Yanyuwa place name, referring to the a reef that is found off the coast of Bing Bong. It is there that a lone ancestral sea turtle travelled to meet with the other sea turtles during the *Yijan*, or ancestral time (Bradley 1997: 503). Here the sea turtle ancestors stayed for some time, singing. Some of these turtles were carrying sacred hollow log coffins (Bradley 1997: 503).

7 Seagrass is a vital resource for the region's dugong populations, and dugong is a Yanyuwa ancestor of the highest order.

References

ABC News. 2015a. *McArthur River Mine: Glencore Share Fall Raises Concerns about Environmental Liability*. Available at www.abc.net.au/news/2015-09-30/glencore-share-falls-raises-concerns-mcarthur-river-mine-bond/6815756. Accessed 10 December 2015.

ABC News. 2015b. *EDO NT Points Finger at McArthur River Mine Over Potential Contamination of More than 400 Cattle*. Available at www.abc.net.au/news/2015-08-22/green-group-blames-mcarthur-river-mine-amid-cattle-lead-scare/6716346. Accessed 13 December 2015.

Aboriginal Land Commissioner. 1979. *Borroloola Land Claim: Report by the Aboriginal Land Commissioner, Mr Justice Toohey, to the Minister for Aboriginal Affairs and to the Minister for the Northern Territory*. Canberra: Australian Government Publishing Services.

Anderson, J. and Monteath, P. 2008. Ernest Favenc and the Exploration of the Northern Territory. *Journal of Northern Territory History* Vol. 19: 1–16.

Australian Indigenous HealthInfoNet. 2015. *Kidney Health*. Available at www.healthinfonet.ecu.edu.au/chronic-conditions/kidney. Accessed 15 December 2015.

Avery, J. and McLaughlin, D. 1977. *Submission by Northern Land Council to Aboriginal Land Commissioner on Behalf of Traditional Aboriginal Owners in Borroloola Region of the Northern Territory*. Unpublished paper, Northern Land Council, Darwin.

Baker, R. 1990. Coming In: The Yanyuwa as a Case Study in the Geography of Contact History, in *Terrible Hard Biscuits*. Edited by V. Chapman and P. Read, pp. 123–166. Sydney, New South Wales: Allen and Unwin.

Baker, R. 1999. *Land Is Life: From Bush to Town, the Story of the Yanyuwa People*. St Leonards, UK: Allen and Unwin.

Baker, R., Davies, J. and Young, E. (eds.). 2001. *Working on Country: Indigenous Environmental Management of Australia's Lands and Coastal Regions*. Melbourne: Oxford University Press.

Bardon, J. 2014. McArthur River Mine's Burning Waste Rock Pile Sparks Health, Environmental Concerns among Gulf of Carpentaria Aboriginal Groups. *ABC News*, 27 July. Available at www.abc.net.au/news/2014-07-27/mcarthur-river-mine-gulf-of-carpentaria-anger-smoke-plume/5625484. Accessed 10 December 2015.

Bardon, J. 2015a. Borroloola Clans Gain More Native Title Rights over NT Cattle Stations. *ABC News*, 26 November. Available at www.abc.net.au/news/2015-11-26/borroloola-clans-gain-more-native-title-rights-cattle-stations/6978540. Accessed 15 December 2015.

Bardon, J. 2015b. McArthur River Mine Negotiations over Environmental Bond 'Good', Glencore says; Green Groups Concerned about 'Cut and Run'. *ABC News*, 24 August. Available at www.abc.net.au/news/2015-08-24/glencore-facing-increased-environmental-bond-at-borroloola/6720056. Accessed 10 December 2015.

Bardon, J. 2015c. NT Chief Minister Threatens McArthur River Mine with Closure Unless it Improves Environmental Practices. *ABC News*, 25 August. Available at www.abc.net.au/news/2015-08-25/nt-chief-minister-threatens-mcarthur-river-mine-with-closure/6724890. Accessed 10 December 2015.

Bauman, T. 2002. 'Test 'im Blood': Subsections and Shame in Katherine. *Anthropological Forum: A Journal of Social Anthropology and Comparative Sociology* Vol. 12(2): 205–220.

Bradley, J. 1988. *Yanyuwa Country: The Yanyuwa People of Borroloola Tell the History of Their Land*. Translated and illustrated by J. Bradley. Melbourne, Victoria: Greenhouse Publications.

Bradley, J. 1997. *Li-Anthawirriyarra, People of the Sea: Yanyuwa Relations with their Maritime Environment*. PhD Dissertation, Northern Territory University, Darwin, Australia.

Bradley, J. 1998. "We Always Look North": Yanyuwa Identity and the Maritime Environment, in *Customary Marine Tenure*. Edited by N. Peterson and B. Rigsby, pp. 125–141. Oceania Monograph. Sydney, New South Wales: University of Sydney Press.

Bradley, J. 2001. Landscapes of the Mind, Landscapes of the Spirit: Negotiating a Sentient Landscape, in *Working on Country: Indigenous Environmental Management of Australia's Lands and Coastal Regions*. Edited by R. Baker, J. Davies and E. Young, pp. 295–304. Melbourne, Victoria: Oxford University Press.

Bradley, J. 2003. Singing Through the Sea: Song, Sea and Emotion. *EarthSong* Autumn 2013: 5–8.

Bradley, J. 2006. The Social, Economic and Historical Construction of Cycad Palms among the Yanyuwa, in *The Social Archaeology of Australian Indigenous Societiesi*. Edited by B. David, B. Barker and I. McNiven, pp. 161–181. Canberra: Australian Studies Press.

Bradley, J. 2008. Singing Through The Sea: Song, Sea and Emotion, in *Deep Blue: Critical Reflections on Nature, Religion and Water*. Edited by S. Shaw and A. Francis, pp. 17–32. Abingdon, Oxon: Routledge.

Bradley, J. 2010. *Singing Saltwater Country*. Sydney, New South Wales: Allen and Unwin.

Bradley, J. 2012. Hearing the Country: Reflexivity as an Intimate Journey into Epistemological Liminalities. *The Australian Journal of Indigenous Education*. Vol. 41(1): 26–33.

Bradley, J. and Kearney, A. 2011. 'He Painted the Law' – William Westall, Stone Monuments and Remembrance of Things Past in the Sir Edward Pellew Islands. *Journal of Material Culture* Vol. 16(1): 25–45.

Bradley, J. and Yanyuwa Families. 2007. *Barni-wardimantha Awara (Don't Spoil the Country): A Yanyuwa Sea Country Plan*. Hobart, Tasmania: National Oceans Office.

Bradley, J. with Yanyuwa Families. 2016. *Wuka nya-nganunga li-Yanyuwa li-Anthawirriyarra: Language for Us, the Yanyuwa Saltwater People: A Yanyuwa Encyclopedic Dictionary*. Melbourne, Victoria: Australian Scholarly Publishing.

Bradley, J. with Kirton, J. and the Yanyuwa Community. 1992. *Yanyuwa Wuka: Language from Yanyuwa Country*. Unpublished document, available at: https://espace.library.uq.edu.au/view/UQ:11306/yanyuwatotal.pdf. Accessed 26 November 2015.

Brown, C. 2014. Top End Anglers Demand Action on Drifting Iron Dust. *ABC Rural News*, 2 September. Available at www.abc.net.au/news/2014-09-02/territory-fishers-call-for-action-on-gulf-port-dust-issues/5710368. Accessed 13 December 2015.

Brown, C. and McCarthy, M. 2014. Plummeting Iron Ore Price Claims Third Top End Miner. *ABC Rural News*, 8 September. Available at www.abc.net.au/news/2014-09-08/western-desert-resources-goes-into-voluntary-administration/5722388. Accessed 13 December 2015.

Burbank, V. 1985. The Mirrirri as Ritualized Aggression. *Oceania* Vol. 56(1): 47–55.

Glencore. 2015. *McArthur River Mine – Shipping*. Available at www.mcarthurrivermine.com.au/EN/AboutUs/Pages/Shipping.aspx. Accessed 2 December 2015.

Hamann, M., Schäuble, C., Simon, T. and Evans, S. 2006. Demographic and Health Parameters of Green Sea Turtles Chelonia Mydas Foraging in the Gulf of Carpentaria, Australia. *Endangered Species Research* Vol. 2: 81–88.

Hamilton, A. 1979. *Timeless Transformations: Women, Men and History in the Western Australian Desert*. PhD thesis, University of Sydney, Sydney, Australia.

Hiatt, L.R. 1978 Classification of the Emotions, in *Australian Aboriginal Concepts*. Edited by L.R. Hiatt, pp. 182–187. Canberra: Australian Institute of Aboriginal Studies.

Howell, E. 2007. *How Do you Put the Rainbow Serpent on the Table? A Discussion of How Knowledge of Place is Contested at the Site of the McArthur River Mine*. Honours thesis, School of Political and Social Enquiry, Monash University.

Howell, E. 2008. The Struggle to Save the McArthur River. *Chain Reaction* Vol. 103: 35.

Kearney, A. 2009a. Homeland Emotion: An Emotional Geography of Heritage and Homeland. *International Journal of Heritage Studies* Vol. 15 (2–3): 209–222.

Kearney, A. 2009b. *Before the Old People and Still Today: An Ethnoarchaeology of Yanyuwa Places and Narratives of Engagement*. North Melbourne: Australian Scholarly Publishing.

Kearney, A. 2010. The Ethnoarchaeology of Engagement: Yanyuwa Places and the Lived Cultural Domain in Northern Australia. *Ethnoarchaeology: Journal of Archaeological, Ethnographic and Experimental Studies* Vol. 2(1): 99–120.

Kearney, A. 2014. *Cultural Wounding, Healing and Emerging Ethnicities: What Happens when the Wounded Survive?* New York: Palgrave Macmillan.

Kearney, A. and Bradley, J. 2015. When a Long Way in a Bark Canoe Becomes a Quick Trip in a Boat: Changing Relationships to Sea Country and Yanyuwa Watercraft Technology. *Quaternary International* Vol. 385: 166–176.

Kirsch, S. 2007. Indigenous Movements and the Risks of Counter Globalization: Tracking the Campaign against Papua New Guinea's Ok Tedi Mine. *American Ethnologist* Vol. 34(2): 303–321.

Kirton, J. and Timothy, N. 1977. Yanyuwa Concepts Relating to 'Skin'. *Oceania* Vol. 47(4): 320–322.

Langton, M. 2002. The Edge of the Sacred, the Edge of Death: Sensual Inscriptions, in *Inscribed Landscapes: Marking and Making Place*. Edited by M. Wilson and B. David, pp. 253–269. Honolulu: University of Hawaii Press.

Latimer, C. 2011. Western Desert looks to Xstrata for Iron Ore Export. Australian Mining. Available at www.australianmining.com.au/news/western-desert-looks-to-xstrata-for-iron-ore-export. Accessed 9 December 2015.

Lehmann, D., Tennant, M., Silva, D., McAullay, D., Lannigan, F., Coates, H. and Stanley, F. 2003. Benefits of Swimming Pools in Two Remote Aboriginal Communities in Western Australia: Intervention Study. *British Medical Journal* Vol. 327: 415–19.

McArthur River Mining. 2007. The River Xstrata Wants to Mine, Part II. Barbara McCarthy. Available at https://mcarthurriver.wordpress.com/2007/02/02/barbara-mccarthy. Accessed 13 December 2015.

McArthur River Mining. 2009. Community Update, in *Memorandum* Edition 10, April. Available at www.mcarthurrivermine.com.au/EN/Publications/Memorandum/090415_MRM_Memorandum%20edition%2010_final.pdf. Accessed 13 December 2015.

McArthur River Mining. 2015. Eyes on the Road, in Memorandum Edition 32, June. Available at www.mcarthurrivermine.com.au/EN/Publications/Memorandum/Memorandum_Edition32.pdf. Accessed 14 December 2015.

Morris, B. 1989. *Domesticating Resistance: The Dhan-gadi Aborigines and the Australian State*. London: Berg.

Muir, H. 2004. *My Very Long Journey: My Life as I Remember It*. Canberra: Australian Institute of Aboriginal and Torres Strait Islander Studies.

Munksgaard, N.C. and Parry, D.L. 2001. Trace Metals, Arsenic and Lead Isotopes in Dissolved and Particulate Phases of North Australian Coastal and Estuarine Seawater. *Marine Chemistry* Vol. 75: 165–184.

Munksgaard, N.C., Brazier, J., Moir, C. and Parry, D. 2003. Trace Metals, Arsenic and Lead Isotopes in Dissolved and Particulate Phases of North Australian Coastal and Estuarine Seawater. *Australian Journal of Chemistry* Vol. 56: 233–238.

Northern Territory Environmental Protection Authority. 2014. *NT EPA Issues Direction to Western Desert Resources to Address Environmental Concerns*. Available at www.ntepa.nt.gov.au/news/2014/WDR-issued-direction. Accessed 13 December 2015.

Pearson, C. 2015. Environmental Concerns for McArthur River. *The Australian Mining Review*. Available at http://australianminingreview.com.au/New/environmental-concerns-for-mcarthur-river. Accessed 13 December 2015.

Polidor, A. and Tindall, A. 2006. McArthur River, *Sacred Land Film Project.* Available at www.sacredland.org/index.php/mcarthur-river. Accessed 9 December 2015.

Povinelli, E. 1993. 'Might Be Something': The Language of Indeterminacy in Australian Aboriginal Land. *Man* Vol. 28(4): 679–704.

Read, P. 1999. *A Rape of the Soul So Profound: The Return of the Stolen Generations.* Sydney, New South Wales: Allen and Unwin.

Roberts, T. 2005. *Frontier Justice: A History of the Gulf Country to 1900.* St Lucia, Queensland: University of Queensland Press.

Roberts, T. 2009. The Brutal Truth: What Happened in the Gulf Country. *The Monthly Essays,* November. Available at www.themonthly.com.au/issue/2009/november/ 1330478364/tony-roberts/brutal-truth. Accessed 15 December 2015.

Rose, D.B. 1992. *Dingo Makes Us Human: Life and Land in an Australian Aboriginal Culture.* Cambridge: Cambridge University Press.

Rose, D.B. 2002. *Country of the Heart: An Indigenous Australian Homeland.* Canberra: Aboriginal Studies Press.

Rose, D.B. 2008. Dreaming Ecology: Beyond the Between. *Religion and Literature* Vol. 40(1): 109–122.

Rose D.B. 2013. Death and Grief in a World of Kin, in *The Handbook of Contemporary Animism.* Edited by G. Harvey, pp. 137–147. Durham, UK: Acumen.

Seton, K. and Bradley, J. 2004. "When You Have No Law You Are Nothing": Cane Toads, Social Consequences and Management Issues. *The Asia Pacific Journal of Anthropology* Vol. 5(3): 205–225.

Spencer, J., Silva, D., Snelling, P. and Hoy, W. 1998. An Epidemic of Renal Failure Among Australian Aboriginals. *Medical Journal of Australia* Vol. 168: 537–541.

Stanner, W.E.H. 1979. *Report on Field Work in North Central and North Australia, 1934–35.* Canberra: Australian Institute of Aboriginal Studies.

Trigger, D. 2008. Indigeneity, Ferality and What 'Belongs' in the Australian Bush: Aboriginal Responses to 'Introduced' Animals and Plants in a Settler-Descendant Society. *Journal of the Royal Anthropological Institute* Vol. 14: 628–646.

Yanyuwa Families, Bradley, J. and Cameron, N. 2003. *Forget About Flinders: An Indigenous Atlas of the Southwest Gulf of Carpentaria.* Canberra: Australian Institute of Aboriginal Studies.

Young, A. 2015. McArthur River Mine: The Making of an Environmental Catastrophe. Available at http://nela.org.au/NELA/NELR/150306_Young_Macarthur_River_ Mine.pdf. Accessed 2 December 2015. Also published as: MRM: What a Mess, *Land Rights News,* January 2015. Available at www.nlc.org.au/files/pdfs/LRN-Jan2015-1-24-A.pdf. Accessed 2 December 2015.

Part III

Diagnosing place harm

Throughout the following three chapters, the emphasis is on diagnosing place harm. This begins with the necessary step of critically examining and delimiting an understanding of what constitutes an 'agent of harm'. It raises the question of how the human as an agent of harm is distinct from others in its ideologies and actions. By way of localised case studies and close readings of harmful moments, which have led to cultural wounding and place decline in a range of cultural contexts, certain qualities are revealed as to the nature of harm and the methods by which it is achieved. How these harms damage place and saturate into the lives and existences of ethnic groups is also explored.

Diagnosing place harm reveals that there are a number of strategies by which the intent to destroy and debilitate both people and place as part of a symbiotic suffering is achieved. These are distinguished as destruction and designification of place, social disordering through renaming and denial of inherent place order, elemental erasure by way of toxicity, contamination and the killing of non-human life. By coming to know these patterns of place harm, it is possible to recognise their presence in contexts all over the world and track back from this awareness to examine the axiologies that support not only cultural wounding, but also its greater effects as violence and trauma in place.

3   Destruction and designification

This chapter, in association with Chapters 4 and 5, serves to illuminate how it is that place becomes wounded. In the preceding chapter, the reader has been called upon to witness the extent of place harm across Yanyuwa homelands in Northern Australia. This ethnography of place harm has been recounted in an effort to not only 'come to know' of violence in place but to generate 'faith in the claim to consequence'. The latter is a more profound realisation of place harm, and one that fully activates the role of witness as one that is fully aware of and even disturbed by the traumatic effects of place harm. The following three chapters now serve to explain what it is that disturbs us as witnesses of place violence and does so by diagnosing and examining the very nature of violence enacted against place and manifest in place harm.

This discussion, however, is not without its challenges. From the outset, the challenge is that place may be wounded in a great number of ways, through human and non-human intervention. As such, I must begin by reflecting on the notion of 'agents of harm', both human and non-human, seeking to understand what distinguishes one from the other and how cultural interpretations may alter the perception of violence and deliberacy in the moment of place harm. Teasing out the wounding effects born of human conflict from those of 'natural disaster' or 'natural calamity', if at all possible, is key to finding the direction for this chapter and subsequent reflections. It is what enables the work to convey the specificity of human interventions in place order that are distinctly harmful. Ultimately, this chapter prioritises a discussion of place wounding as brought about by human conflict and deliberate human violence, seeking to reveal the kinds of conflict and violence that bring about uncertainty and rapid, detrimental change. That violence in place occurs is attested to by a substantial archive of recorded and living memory. This reveals the possibility of place harm and its ongoing effects on human life and esteem. The aim here is to examine some of this material, identifying themes to the experience of place wounding. Yet, before doing so, it is important to reflect on why the exercise of untangling the human from the non-human as agent of harm is difficult in the context of this book. The primary challenge stems from the methodological orientation of this work, in particular its orientation towards an Indigenous epistemology of place and kincentric ecology.

Agents of harm

Where distinctions between nature and culture move beyond the binary of post-Enlightenment rationalism, separating human from non-human action and agency becomes difficult. The challenge is no more vividly conveyed than through reflection on what constitutes an agent of harm; particularly when that agent is referred to as a 'natural disaster'. With 'natural disasters', the encounter is often considered cataclysmic and through seemingly sudden events the tangible quality of place is impacted, precipitating substantive rearrangements and disorder in place. These disasters can trigger shifts in the bio and geo spheres. They may also cause shifts to occur within the intellectual or social spheres that hold place and render it meaningful in relation to human life. This sphere of cognition is the rapidly expanding realm of human thought and place consciousness, which allows people to make sense of place as it once was, prior to harm, and as it may come to be in the aftermath of disaster. Disorder in place may result in abandonment, or a reinscription of place as a non-place, thus the relational meanings that are created through the people and place encounter shifts in the aftermath of disaster. These reflect the ways in which interacting forces run between people and place, implicating both as agents of influence. It is often held that people construct meaning in and of place, yet place, having its own order and inherent agency, is equally vested in the construction of meaning, both before and after a disordering event.

Parallels may be drawn in discussions of post-traumatic and related stress disorders that emerge from horrific events, as triggered by both 'natural' and human agents of harm. Kai Erikson (1976, 1994, 2011), an authority on the social consequences of catastrophes, has written extensively on the devastating personal and communal effects that disasters and place disorder have on people. In his 1994 work, *A New Species of Trouble: Explorations in Disaster, Trauma and Community*, he distinguishes the human as the new species of trouble, the great agent of harm. Identifying a particular experience of trauma and violence, human-induced disasters – a broken dam, methyl mercury poisoning, radiation leakage – are, according to Erikson (1994) unlike natural disasters: they have preventable causes. Because of this, "there is always a story to be told about them, always a moral to be drawn from them, always a share of blame to be assigned" (Erikson 1994: 142). As a result of these events, collective trauma sets in, and according to Erikson, 'I' continues to exist, 'you' continues to exist, although distant and hard to relate to. But 'we' no longer exist as a connected pair or as linked cells in a larger communal body (Erikson 1976: 154). The disordering of place by this species of trouble (the human) thus has a consequence on social life, may degrade community relations, rupture a sense of justice by compromising the rights and equality of a few, but not all. The experience of 'natural disaster' has a different effect on the "we" of the communal body. The communal body of human life (irrespective of difference) is cast as the recipient of 'nature's' great force. Rather than fracturing the human whole that is community, this can instead strengthen bonds and instate wider networks of empathic response and understanding, as seen in the giving of foreign aid and international support with disastrous events such as the Fukushima disaster in 2011 and the Boxing Day tsunami of 2004.

Sociologists and cultural geographers alike (Barton 1969; Blaikie et al. 1994; Dyson 2006; Eyerman 2015; Hannigan 2012; Peacock et al. 1997; Pelling 2003) have written of the trauma that comes to reside in place with the effects of natural disasters, and the forthcoming struggle to survive. These accounts differ from narratives of place destruction caused by human-induced disasters such as war and ethnic violence. Although many authors recognise the rapidity with which 'natural disasters' transition into social ones (see Eyerman 2015) and reflect the culturally prescribed ways in which governments, communities and individuals attempt to deal with and survive catastrophe, most write from a Western perspective, and assume some universalism in the human experience and conceptualisation of 'nature' as an actual agent of harm. Nature as opponent is a greater-than-human life presence. It is 'nature' and it is 'nature's fury' to which we are exposed, considered awesome in its destructive power. Such 'nature' events are taken to remind us that we, as humans, along with other presences in the biosphere, are small and vulnerable, and that living on this dynamic planet always entails risk. The enormity conveyed by this opponent relies upon the axiological and ontological shrinking of human life, and acceptance of a place for the human within a 'natural world'.

Death and dispossession may be explained as 'inevitable or expected losses' in the wake of nature's fury (Nicosia 2009: 1). However, when the perpetrator of violence against place and its people is an ethnic other, the state or neoliberal market forces and their operators, and the trauma is located within a cultural context, comfort is not found in accepting a relational smallness that makes the harm of place and its people inevitable, expected or rational. Rarely do those who have been subject to ethnic violence, loss of home and place annihilation, chalk it up to an inevitability that befalls the weak and benefits the strong. While it is 'innocent victims' who loose lives in terrorist attacks, or civil war, a rhetoric of 'naturalness' is employed when human fatalities are a 'consequence' of 'natural disaster' (Nicosia 2009: 2). The 'natural world' is, in the case of a 'natural disaster', seen to hold little regard for the cultural universe of human existence. It operates according to an inner logic that human life is compelled – however baffling it might be – to accept.

In the aftermath of natural disaster, people do not call upon nature to recompense, nor do they seek retaliation or retribution, as they might when faced with the enduring suffering that comes from place destruction that is deliberately enacted by powerful ethnic others. While there is no doubt that events triggered by natural disasters rearrange place, fracturing the kinship that binds people to sites of significance, there is evidence to suggest that they are perceived in such a way as to ensure they can be accommodated into living memory and also a rhetoric of survival. In many instances, profound and harmful shifts in the health and well-being of human populations are traced to place decline following disastrous events (see Alexander 1999; Cherry 2009), yet a sense of blame is often diffused. Despite this, their impact is very real and requires epistemological structures to process, along with ontological and axiological counsel to translate their meaning into a functioning sense of survival and everyday life. That said, how traumatic events, which reorder the place world, are rendered meaningful or understandable and thus accommodated through survival and renewal depends on the cultural universe in which people coexist with place. Given there are distinctively different ways of knowing the place world and appreciating the depth of its agency, as outlined in Chapter 1, I am left asking how else 'disasters' of the 'natural or unnatural' kind might be thought of as enacted by agents of harm.

The concept of a 'natural disaster' and the binary it infers between 'nature' and 'culture' is not universal. It cannot be produced nor upheld by an epistemological framework in which the world is conceived of as deeply cultural, and made up of multifarious agents and co-presences (as discussed in Chapter 1). Such is the case for Indigenous epistemologies. For many Indigenous groups, the disastrous effects of cyclones, floods or prolonged periods of drought can be attributed to the actions of ancestral beings or those who ensorcell place and people with harm. This is traceable to the fact that weather is a moral agent and thus has agency. Weather events might also be read as ancestral responses to disorder in place, a form of aggressive reminder that human kin must take up responsibility for the place world and all it contains, reinstating order through kinship and obligation. Disordering events such as flood or species decline may be considered retributive acts on the behalf of place itself or the ancestors that control it. Disorder becomes part of a communicative event through which human life is called to witness and respond.

What are often referred to in geoscience as 'rapid onset natural hazards', including bushfires, tropical cyclones, earthquakes, floods, landslides, severe storms and tsunami, are widely understood as something else by Indigenous groups throughout Australia and beyond. While much is written on Indigenous knowledge as an instrument to predict impending rapid onset natural disasters as a form of disaster risk reduction, less is documented on the actual Indigenous knowledges that relate to causation and origins of such events (Baumwoll 2008; Mercer et al. 2009; Shaw et al. 2009). Some have documented Indigenous knowledge of weather and explore weather as a moral agent, yet it would seem the latter remains an oversight in the existing literature (Bradley 1997; see Green et al. 2010; Green and Raygorodetsky 2010; Orlove et al. 2010). A possible reflection of epistemological gaps in dominant pedagogy of place and 'nature', this oversight is offset by a few recorded instances, and certainly anecdotal evidence that has come from ethnographic accounts in working with Indigenous families in northern Australia (see Bradley 1997, 2010).

For Yanyuwa, the Indigenous owners of land and sea territories in northern Australia, the ancestors, which occupy an unseen realm in the present, are known to precipitate high floodwaters, tropical cyclones and severe storms as a way of communicating disorder or dissent. Ancestral beings that occupy homelands as an invisible presence for the most part are also known to hide food resources from those hunting and gathering, they allow thickets of weeds to grow out of control and stymy species numbers, in an exercise of 'shutting up' or 'closing off country' (Kearney ethnographic fieldnotes, 2001–2015). This is often done in response to foreign presences on country, such as non-Indigenous persons and mining companies and their extractive technologies, and is a way of communicating a sense of neglect for country, and a breach of ancestral Law.

Drawing on 35 years of ethnography, anthropologist John Bradley recalls Indigenous knowledge of weather events and rapid onset disasters, as follows,

> Cyclone Tracy, which occurred in Darwin, northern Australia in 1974, was triggered by a sacred stone, which had been removed from a very important reef just off the coast. Cyclone Kathy, which hit the Gulf of Carpentaria in 1984, was the result of young male initiates not being ritually 'finished off' during their circumcision rites [meaning ceremony was not fully closed at its completion, by way of song and dance elements]. The strength of Cyclone Kathy was also exacerbated by the presence of too many tourists going where they should not on the islands of Yanyuwa country. For the Yanyuwa huge floods are the direct result of anger among senior rain making men, and of the Flood Dreaming site being tampered with in some way.
>
> (John Bradley pers. comm., 11 November 2015)

Each of these events reflects a disregard for Law, leading to an initial disordering of place and a subsequent place response. The capacity to understand these responses as the actions of moral agents is highly developed within this Indigenous worldview, as is the openness to communicative happenings that might tell of disorder and impeding consequence. Betty Pearce, an eastern Arrernte woman, recalls how in the days before Cyclone Tracy (which hit the city of Darwin in the summer of 1974 with devastating consequences), she travelled east to Arnhem Land and the township of Oenpelli (Gunbalanya) in northern Australia. Travelling with local Indigenous persons, she was told that the people of Oenpelli had a warning to give to people living in the Aboriginal community of Bagot, a major Indigenous settlement located close to the centre of Darwin, Australia's most northern city (where the cyclone would do its greatest damage) (in Liston and La Canna 2014; see also Cunningham 2014 for an account of events). She says, upon visiting a rocky place in Oenpelli, with Indigenous owners and family she "realised there were snakes coiled up and sunning themselves and ... goannas lying only a few feet away, a metre or so away from each other". "Goannas were facing the sun, the morning sun, but there were other goannas with their backs to the sun facing the west" (Liston and La Canna 2014). Of this, she recalls saying, "Ooh that's funny, I've never seen that before", to which her hosts replied, "No, this is why we're telling you, you tell everybody to come back. Everybody you see you tell them to come back to their country because something really bad's going to happen but we don't know what it is. Snakes and goannas – they're natural enemies and here they are sunning themselves on the same slabs of rock" (Liston and La Canna 2014).

Where it is felt that 'things are not right', disastrous or untimely events may occur. For Yanyuwa, these are read as revelations that the 'old people' are not happy and are responding to disorder or neglect in place. In recent decades, it has been alien or feral presences, human and non-human, which are held overwhelmingly responsible for the 'rubbishing of country', an expression taken to mean harm caused in place as a result of disregarding place law and importance (Kearney 2009). The ancestors may also act out towards Yanyuwa kin who break the law of place. Transgressing rules of access and clan boundaries has been cited as grounds on which ancestors may affect species numbers, or cause rivers to flood, and interrupt the overall health of place and its elemental parts (Kearney 2009; Kearney and Bradley 2009; Bradley and Kearney 2011; Bradley 1997, 2000, 2001, 2010). Malevolence of this kind is strongly balanced by deep benevolence in the case of Yanyuwa kin, yet is necessary as a form of interruption where systemic disorder brought on by other human agents (Indigenous owners or otherwise) impacts on place, its Law and its health. Illustrative of disorder, catastrophic and disordering moments, which the West refer to as 'natural disasters' are communicative events in which co-presences convey the experience of harm in place and place disorder. There is nothing natural in their character or occurrence, nor do people tend to find themselves adrift for answers as to why these events occur. There are sophisticated epistemological and ontological structures available for those seeking to understand these events and what they communicate.

It is apparent that while some epistemologies attribute the cause of harm to 'nature', for others it is the work of weather as a moral agent, ancestral beings and overwhelmingly powerful agents or co-presences that reside in place or are place itself. These co-presences may be spirit men and women, and other ancestral spirits seeking to educate their human counterparts. Rainbow serpents, coyotes, caribou and reindeer, rain birds, deer or salmon are just a few mentionable co-presences capable of agency, whether conveyed through acts of kindness or aggression (Anderson 2004; Fienup-Riordan 1994; Trigger 2008). These agents relate to humans in culturally prescribed ways and often perform a key role in the making of place, the sustaining of its order and revelation of its disorder. From a non-Indigenous perspective, while harm-doing might be attributed to 'nature', the 'nature agent' is still, although to a lesser extent, read as having agency, and is granted a personal and/or place-based identity, often through naming and the attribution of personality. Hurricane Katrina and Cyclone Kathy remind us that these events were held to have specific characters, they were charged with indiscriminate cruelty and through naming are brought into a knowable human realm. While this hints at the possibility of the West acknowledging otherly co-presences, non-Indigenous epistemologies rarely carry the same relational depth of interconnection and kinship that Indigenous place knowledge is capable of.

In the case of 'natural' disasters, these 'agents' or powerful co-presences are held responsible for the substantive rearrangement of place. They are, however, for the most part, classified as distinct from the human agents that seek to enact harm over place. For example, Yanyuwa do not classify ancestral agents, or malevolent forces in and of country, as they would the colonial presences that have plundered their homelands, rendered the lands and waters ill, and driven down the strength of people and place in their relentless pursuit of land and resources. Perhaps what distinguishes one from the other is that in the first instance, while there is intent to harm, it is of a different kind. The desire to erase certain ethnic and cultural presences and cause enduring suffering is not within the remit or realm of ancestral presences and powerful co-presences in place that recognise kinship. The communicative intention of their actions and their interventions in place are something different and reflect the ordering and aligning logic of kinship and mutual responsibility, obligation and reciprocity.

In the case of Indigenous Australian experiences, erasure and enduring suffering has, however, been the motivation of those who sought to advance colonial expansion and its sister project of modernity at the cost of people and place health (for a critical discussion of modernity as coloniality, see Mignolo 2007). Harmful intent is directed specifically at human presences in place, as an obstacle to the colonial and developing neoliberal project, and place becomes the object of desire. This is unlike the effects of change brought about by powerful co-presences (such as ancestral beings) or non-human agents in place, which may perform the role of recalibrating place, reordering relationships between people and place, or emphasising a new era in the biography of place and people. What distinguishes human action as the catalyst for violence and wounding, as witnessed with colonial expansion and modernity as neoliberal resource industries and 'developmental' agendas, is first, the intent to harm place and its human kin, and second, a disavowal of place law and existing order.

It is the denial of existing order both in the ecology and integrity or law of place and the kinship that aligns people with place that leads to substantial harm in the place world. Separation of the cultural realm, from the 'natural' one is what fuels the axiological crises that allow violence to occur in place as a greater effect of cultural wounding. Axiology refers to the perception of worth and sense of value as it is generated within a cultural context. What people come to value and how they determine wealth is culturally prescribed, thus there are distinctions to be found in the way that people perceive and prioritise place. Where one cultural group determines value, in this case through kinship and recognised agency in place, another may determine zero value or a value of economic or extractive quality which is arrived at only through the expulsion of human presences. The crisis comes when the latter overrides the former and value is denied or slips from human consciousness. It also lies in the failure or inability to find pathways towards kinship, motivations to care and the capacity to see place value.

Henceforth, the illustrative accounts of place harm that are described in this chapter and the two that follow involve violent and wounding moments made possible because of the denial of place order, character and kinship. Articulated in three parts – first as place destruction and designification (Chapter 3); second, social disorder in place (Chapter 4); and lastly, ecological decline, toxicity and elemental decay (Chapter 5) – it is argued that violence and disorder, which lead to wounding, are achievable through the denial of kinship, place agency and inherent place order. It is proposed that there is a difference between 'natural disasters' as powerful events, whether attributed to 'nature' agents or powerful co-presences, and human actions as intent laden and deliberately violent in the service of cultural wounding. In each treatment of place harm, an account of violence is presented. These accounts recall a pivotal moment of violence in place. As close readings of place harm and wounding, they set the scene for a critical reflection on types of harm and the axiological crises that enable them. It introduces the very real and lived experiences of loss and suffering that have befallen place and, as a result, its people. These are stories of tragedy in beautiful places, places that inherently matter and exist in relation to human kin.

Destruction and designification of place

I begin this diagnosis of place harm by considering destruction and designification. In this instance, the focus is on the physicality of place. By examining the wounding of place through violent acts such as annihilation, bombing, large-scale resource projects, site desecration and ethnic cleansing, place is returned to its spatial roots. This is in recognition of the tangible quality of the place world in our everyday lives. Part of our haptic geography and sensory capacity, place affects our corporeal existence and in turn human presence triggers recognition and response in place, whether through shifts in air quality, rising temperatures, air pressure or smell, or even more aggressively through bombing and destruction. As people and place are inseparable and mutually sentient, cultural wounding and its violent mechanisms can lead to root shock for both. Root shock of this kind is palpable, and derives from acts of destruction and designification (Fullilove 2013). It registers in the corpus of both people and place.

Place destruction and designification are recalled through human accounts of loss. They are evidenced by disappearing landscapes, scarred cites, empty villages and those places that have been torn by violent acts such as bombing, nuclear testing and large-scale resource projects. While not an exhaustive list, these examples alone provide much scope for an evaluation of the ways in which place is destroyed and/or designified. Destruction as an act is often a wanton one. Wanton in the fact that it is unprovoked, gratuitous, capricious and unjust. Little is left behind when the intent to destroy is met, and if complete physical erasure is not achieved, then partial destructions of place can lead to a deidentification of place or loss of core place-meaning. Hewitt (1983: 257–258) writes that this kind of violence "intends the disorganization of enemy space". When destruction is not sought, and an essence of place physicality must remain to meet a new purpose in place use, then designification is an alternative that ensures confusion in place and the denial of any existing order and value.

Destruction and designification are conspiring themes in the effort to wound place, one bringing about physical chaos (along a spectrum from minimal impact to a final death point) in place, the other a denial of this impending chaos or any existing merit and order, which becomes silenced through an overlaying of new meaning. Removing any explicit reference to place, 'destruction' and 'designification' have menacing etymologies frequently used to reference acts of violence and dismissal. By extension, these meanings take on decisive qualities of threat when the object of attention is place. Drawing on the definitional limits of the verbs 'to destroy' and 'to designify', the following come to light:

> To destroy: end the existence of something by damaging or attacking it, to ruin someone or something emotionally or spiritually, to defeat utterly. To crush, main, ravage, ruin, deface, erase, extinguish, annihilate, mutilate, nullify, slay, waste, or tear down.

> To designify: conceal, withhold, bottle up, cover, hide, leave alone, refrain, refuse, supress. To bash, belittle, berate, cut down, insult, minimise, persecute, put down, revile, or vilify.

It is designification that supports a denial of place order, and disavowal of existing kinship. Denying the character of place is necessary for what Falah (1996) terms resignification of place. Resignification is prefaced by the knowledge that an ideological sediment exists in place. These sedimentary layers of meaning, order and value are deposited by human presences, or self-generated by place and its ancestral makers. Once identified, they can be denied, or erased, and then overlain with the order of another (Falah 1996: 256; see also Schnell and Mishal 2008). Designifying and resignifying efforts are known to have occurred across Palestinian villages, during and after the 1948 Israeli–Palestinian War (Falah 1996; see also Braverman 2008; Khalidi and Elmusa 1992; Parmenter 1994). Some 418 Palestinian villages endured the force of total war, while human counterparts faced expulsion and prevention of return (Falah 1996: 261). Abufarha (2008) and Falah (1996) write of the transformation of Palestinian cultural landscapes, painting an historical geography of a landscape's designification and resignification.

The act of designification is presented as one aimed at uprooting people from place and then obliterating the vacated places (Falah 1996: 257). Whether human kin were expelled or became part of an exodus, the first wave of shift in place character came with the absence of a certain human presence, one that was/is ethnically and culturally prescribed. "Space purification" being an apt description for the removal as it took place (Sibley 1988). Described also as ethnic cleansing, removing people worked to undercut and weaken Palestinian claims to territory (Falah 1996: 257) and in turn satisfied the aim of weakening the identity of place. Yet in turn, as Abufarha (2008: 344) writes, "the most prominent aspect of Palestinian's representations of nationhood and peoplehood across time since their encounter with the Zionist project in Palestine around the turn of the twentieth century has been the articulation of their rootedness in the land of Palestine".

Tracking the historiography of the villages "left behind" after the removal of human habitants, Falah, in consultation with historical records, reveals complete destruction of villages by burning to the ground, bombing and blowing up of debris, vandalism, demolition and a sequence of destructions that completely obliterated homes, left some as rubble, others with walls, some with tattered roofs (Falah 1996: 264). Of the original 418 villages, 407 were revisited by Falah and explored for the long-term effects of destruction (Falah 1996: 268). The remaining 11 lay out of reach, now inscribed as restricted military zones (Falah 1996: 268). Some of the visited villages had experienced absolute erasure; nothing left standing as testimony to a prior existence. Others had been covered by dense and deliberate reforestation, removed from the visual field altogether, with place character changed also through radical shifts in land use, conversion into citrus plantations and crop fields, and even garbage dumps. Elsewhere abandoned olive groves stand as reminders of a prior place identity (Falah 1996: 268).

For those places that avoided destruction, homes have been renovated to become homely abodes in Jewish settlements. The register of destruction contains a complicated ledger of lesser and greater physical impacts, yet the effects remain the same; destructions and designifications of place, delivered with the intent to harm place in order to harm people. Neglect and indifference serve a greater purpose in place wounding, feeding the decay and dereliction of sacred sites, and places of worship. Reverting places of worship to storehouses for animal fodder reinscribes meaning through disavowal of place order and value. Existing character and axiological merit are denied (Falah 1996: 279). Braverman (2008) reflects on the use of the landscape as the 'natural alibi' for conflict. She recalls that "[s]ince 1901, the Jewish National Fund (JNF), an organization established by the Fifth Zionist Congress to purchase land in biblical Israel, has planted more than 240 million trees (most pines) in Israel" (Braverman 2008: 450). "The pine has become the quintessential symbol of the Zionist project of afforesting the Holy Land into a European-looking landscape" (Braverman 2008: 450; see also Amir and Rechtman 2006: 39; Bardenstein 1999; Cohen 1993; Long 2005; Zerubavel 1995). In response, the olive tree has become the symbol of Palestinian resistance as a "steadfast connection (*tsumud*) to the land" (Braverman 2008: 450).

The planting of pine trees, at the cost of olive groves and other landscape elements, illustrates the dual processes of designifying and resignifying, designed to obliterate existing place kinship and order. It is as Braverman (2008: 450) writes, "the pitting of the pine tree/people against olive tree/ people reflects the discursive and material split constructed with much fervency and determinacy by the two national ideologies that compete in and over Israel/Palestine". These trees, as elemental parts of place, have become totemic displacements in the occupied territories. Braverman (2008: 450) is astute in her assessment that the Israeli–Palestinian war is deflected onto the landscape, thus pitting place as the 'natural alibi' for cultural wounding or 'symbol of strength' in the face of harm. Abufarha (2008: 346) extends this richly, exploring place and place elements as articulations of belonging, experience and identity, defiant in the face of incoming agents, as illustrated by the

> *saber* (cactus) as a metaphor of community in pre-1948 Palestine; *al-burtuqal* (the orange) as a symbol of loss and robbed nationhood in the 1950s and 1960s; *al-zaytouna* (the olive tree) as a symbol of rootedness that came to the fore in the 1980s and persisted since as a dominant symbol; the *shakiq* (the poppy) as a symbol of sacrifice in the Palestinian resistance; and the re-emergence of the *saber* in the 1990s in a new form as a symbol of the newly constructed Palestinian identity – defiant, resilient, and resistant, powered by its fusion in its history and the land.
>
> (Abufarha 2008: 346)

Place harbours the instruments of conflict as well as resistance, and thus is made instrumental to the fight by way of physical and ideological designification, resignification, resistance and survival.

Destruction and designification simultaneously strive to rupture the kinship between people and place, achieving their wounds by annihilating a place and removing it from the tangible world of possibilities or denying its existence and reinscribing its value. This may be complete or partial, but as with the wounding of a body, damage to one part of place constitutes damage to the whole. Ill-health compromises the opportunities for place kin to return, relate to and witness place through its survival and replenishment. A physical attack on place severely impacts on the ideological conditions that ensure its survival. The relational impact is evident in accounts of bombings and five years of sustained attacks on the city of Aleppo, Syria or in the annihilation of Beit Hanoun, a Gaza neighbourhood, largely destroyed over the course of one hour on 29 July 2014. For the Syrian locals and the residents of Beit Hanoun, they mournfully remember their homes and the landscape of their ancestors. The loss is etched in memory and in the physicality of place, but no longer as a presence, instead as an absence or as rubble. The flattened scape that was once shaped by homes, heritage and livelihoods is now what triggers utter "despair" for the kin that are left to mourn (Porter 2015).

The importance of kinship between people and place is reiterated through these reflections and others of coming to terms with the loss of home, territory and the places to which identities are bound and find their origins. This has been vividly cast as the narrative of loss and land theft told by Indigenous Australians throughout my 15 years of ethnography in northern Australia, and yet is a story I have also heard told by African descendants living within African Brazilian communities in Brazil, and Greek Cypriots in southern Cyprus. In each instance there is an emotional geography that ties people to place, as a former, now alienated homeland, an ancestral origin, occupied territory or impoverished community defined by experiences of enduring hardship precipitated by cultural wounding. Whether place remains in a physical sense has a huge impact on people's ability to anchor identity, and to utilise that anchorage point as a source of strength and resilience. The extent to which the sedimentary layers of place-meaning have been eroded or denied or even supplanted with another order also has a bearing on how people presently live their lives and the extent to which they might rally together through experiences of hardship. For many, the heartfelt longing to return home and remember home is what defines daily life.

The list of historical and contemporary place destructions is a lengthy one, and designifications have come to characterise the place biographies of all nations. From the felling of sacred trees on Indigenous homelands in northern Australia, to the historical destruction of Jewish communities in Poland, the register is vast and unsettling, detailing the outcomes of culture clashes, ethnic violence, racism and the value nihilism of axiological crises. Destructions in times of war, from Jericho, Carthage, Warsaw, Dresden, Hiroshima and Nagasaki, are matched by those in times of peace, as the modern era's large-scale infrastructure projects and obsession with mining and resource extraction reaches grave new heights. The place world is bound by the experience of destruction and designification at the hands of human perpetrators. The disregard for place order and value is unmatched but for the disregard for some forms of human life. There is no beauty in violence, and the swiftest delivery of harm is achieved through place destruction, purely because place is the source of human meaning and identity, it is the site that provides sanctuary and often justifies the existence of an ethnic group and its ancestry. Without such territories, it is harder to maintain esteem, to legitimate presence, and even retain good health among ethnic constituents. By design, place destruction and disorder initiate and then compound human suffering. The following account of violence and place harm grounds this discussion and presses us to consider how place and its people experience these events simultaneously.

Cheslatta T'En and the drowning of place

In 1952, the Cheslatta T'En, members of the Carrier First Nations, British Columbia, were displaced from their homes and ancestral territories. The waters of the Kenney Dam Reservoir would come to submerge large areas of their homelands. Beginning in 1951 and lasting until 1967, Alcan Aluminium (now Rio Tinto Alcan) undertook the large-scale river diversion project along the Nechako River, a megaproject that saw the flooding of the Nechako River Valley (Windsor and McVey 2005: 146). Involuntarily moved on from their homes, given ten days' official notice of the impending flooding and recompensed a paltry fee per hectare of land, the places of their ancestors and the hub of their livelihood faced its death.[1] The preparations for this event would see Cheslatta T'En homes and the local church razed, ranches bulldozed and then native reservations and graveyards engulfed by water and sediment (Windsor and McVey 2005: 153–155).

Construction involved more than 3 million cubic metres of material. Concrete was injected into bedrock fissures below the dam to stabilise the foundation. Engineers had to first de-water the part of the river valley in which they wished to place the dam. This was achieved by diverting the river through a tunnel. For the Kenney Dam, a 16-kilometre-long tunnel was blasted through the Coast Mountains, connecting the Nechako Reservoir to a hydroelectric powerhouse (Beck 2010: 55–59). The water of the reservoir filled an area of 92,000 hectares and involved flooding 32,000 acres of land (Christensen 1995: 38). The reservoir disrupted existing ecosystems and impacted animals such as beavers, Canadian geese, ducks, moose and about 100 trumpeter swans, equivalent to one-eighth of Canada's swan population (Christensen 1995: 51–52). The impoundment behind the dam forms the Nechako Reservoir, which took four years to fill. During these four years, there was no flow running from the Nechako Reservoir for 50 miles between the dam and the Nautley River (Christensen 1995: 42). The fish left in this section of the river remained trapped and died. Local animals that relied on that section of the river were also forced to migrate (Christensen 1995: 42). In the first two years of operation, smelter emissions from the new aluminium plant were believed to have caused further forest depletion, health problems for local workers and harm to aquatic ecosystems (Environment Canada 2004). Water released from the reservoir scoured the area between Skins Lake and Cheslatta Lake. Soil, gravel, grass, moss, shrubs and trees were washed away (see Wood 2013; Beck 2010).

Despite promises that Cheslatta T'En graveyards would be safeguarded, erosion caused by high waters led to several coffins being carried downstream in the current (Windsor and McVey 2005: 155). A deracinated people, they endured the loss of their place kin as present in ancestral beings, living non-human species and landscape features, and place suffered a drowning that caused catastrophic disorder. Place was lost, not in war, but in a time of alleged peace, as its elemental parts were drowned, and its order submerged to make way for an incoming and alien place order. The deep colonising of the settler colonial state rendered this event possible. It is said that the spiritual and social destruction of the Cheslatta T'En began with the dam's construction, as the health of people declined in direct relation to the health of place (Windsor and McVey 2005: 156).

Relocated onto tracts of land distinguished as the ancestral territories of the neighbouring Wetsuet'en, Cheslatta T'En have struggled to find themselves and kinship in a place ordered for and by others. The telling signs of deep wounding, in the aftermath of place destruction are documented by Windsor and McVey (2005: 156; see also Shkilnyk 1985; Borup et al. 1979; Good 1996) as alcohol and substance abuse, spiritual decline and suicide, premature deaths, sickness and depression. Death by drowning has occurred not just for place, but also, according to Windsor and McVey (2005), for its human kin. Cherished reminders of home, pathways, fields and sites of mourning have all gone, and now lie buried beneath meters of sediment and water (Windsor and McVey 2005:156). Lost too are the species once hunted and gathered in the reciprocal ecology of place. The soil, trees, blades of grass, the birdlife, many mammals and the ancestral tracks in place are gone. These elemental specks of place identity cannot be brought back to life.

In May 2015, human remains from a Cheslatta Carrier First Nation cemetery again washed up on the shores of British Columbia's Nechako River as floodwaters rose (CBC News 2015a, 2015b). Mike Robertson, the nation's senior policy advisor, says the social costs of the flood continue to be high for the small community: "There's the cost of … finding the leg of your grandmother, or your uncle, or your cousin. I've got bones sitting on my desk right now […] When people go down there and they find their relatives, and these aren't ancient graves, these are graves as late as 1952, people are still alive that are directly related. They don't know who these people are" (CBC News 2015b). Media reports, and interviews posted online reveal that Cheslatta Carrier sentiment is low and matched by a deep sadness as to the harm that has been done to this place (see CBC News 2015b).

Sadly not alone in this experience, Cheslatta T'En share these effects of place wounding with Indigenous nations and marginalised populations worldwide, leading to the adage "dam a river, damn a people" (Paine 1982; see also Bartolome et al. 2000; Colchester 2000; Fisher 1999; Gray 1996; McCully 1996; Patridge 1989). Destruction of home through large-scale infrastructure projects, along with mining projects threatens Indigenous homelands and autonomy across the global south and north (Bebbington 2012; Downing 1996, 2002; Gedicks 2001; Omeje 2008; Scambary 2013). Megaprojects, including dams, bridges, tunnels, railways, hydroelectric facilities or nuclear power plants, effect monumental harm and are violent in their capacity to destroy and designify place. For the Cheslatta T'En, the perpetrator of violence is the colonial state. As Porteous (1990:190, cited in Windsor and McVey 2005: 149) reflects: "The heart of the matter is that post-war Canadians have become arrant modernizers, creators of placeless deathscapes, destroyers of wilderness in pursuit of profit."

In fear of a similar fate, the Indigenous nations of the Xingu River Region, in Brazil, including Kayapó, Arara, Juruna, Araweté, Xikrin, Asurini and Parakanã Indians and others dependent on the riverine environment (Ribeirinhos), continue to battle the Brazilian government and halt the intent to finish the construction of the world's third largest hydroelectric dam (Hall and Branford 2012; Pontes Júnior and Beltrão 2004). The battle began in the late 1970s and became a consolidated effort in the 1980s as repeated attempts to launch the development failed, resulting in repeated lawsuits and the most recent Supreme Federal Court ruling (2012) to recommence construction. The Belo Monte Dam Complex, referred to as a 'mega dam', is located on the Xingu River, a major Amazon tributary. The Xingu River is home to "37 distinct ethnicities and 800 fish species", although diversity of fish species is thought to be closer to 4,000 (Pontes Júnior and Beltrão 2004: 28). Now well over 50 per cent complete, the dam complex is designed to divert 80 per cent of the Xingu River's flow and in turn will devastate an area of more than 1,500 square kilometres. The reservoir will cause the direct flooding of more than 400 square kilometres of Brazilian rainforest and trigger the forced displacement of between 20,000 and 40,000 people (Amazon Watch nd; Fearnside 2006; Petras 2013). In conjunction with the secondary dam, the Altamira, the impact will be felt across approximately 6,500 square kilometres of rainforest (Fearnside 2006).

While there has been objection to the development from the international community, and questions as to the project's viability, the axiological premise that sustains its very proposal and likely undertaking is of primary concern here. The impact will be felt across all elements of place, physically it will be destroyed or at least heavily designified of any existing order, ecologically there will be elemental erasure and compromise that comes with species loss, poor water quality and loss of vegetation. Socially and ideologically it is the powerful co-presences that exist in ancestral form, and non-human life form that will be disordered in the coming sequence of place annihilation. Hewitt (1983), building on the work of Bunge (1973), regards annihilation as having its greatest effect on the continuing biological, psychosocial and cultural foundations of human geography. Ancestral markers in place, archaeological sites, hunting pathways, ceremonial sites and the communicative events that inform Indigenous economies will be destroyed or alienated through land tenure arrangements favouring Norte Energia, a consortium controlled by the state-owned power company Eletrobras. That kinship will be sustained between Indigenous people and place under such conditions is highly unlikely. If it does persist, it is likely to take on a new form, not dissimilar to the emerging forms of land and sea association that come about for those Indigenous nations physically alienated from their homelands, yet sustaining in a powerful defence of possession of place spirit (see Kearney 2007).

Between damming and gold mining, Indigenous lands throughout Brazil have been disastrously affected (see Ramos 1996). The destructions and designifications that have been witnessed in Brazil, and elsewhere, find their roots in axiological crises that reveal an abiding problem in 'how to be, and what life is worth living and preserving'. Axiological crises that support the destruction of place, ecosystems, and the kinship that governs place law, "originate primarily from cultural causes, such as the assumed incompatibility of two or more cultural systems, or conflicting social interests" (Bidney 1996: 355). "In extreme instances an axiological crisis may be a reflection of value nihilism, of the anarchic cultural state in which there are no established, socially recognized standards, and ... parties act without scruples and with a view solely to their own immediate advantage" (Bidney 1996: 355). That these places contain richness and order for their human kin is incompatible to both the ontology and axiology of mega dams and other such developments. This incompatibility must be sidestepped or reduced to nothing in order for place destruction to occur.

Writing on themes of ecology and violence, Heider (2005: 9) reflects "in destruction, one replaces an order with a lower degree of order, or lack of order [disorder]". Destruction entails transformation to nothing (Heider 2005: 9–10), or equally the ruin of power and function, and a breaking into pieces (Heider 2005: 10). In distinguishing order from disorder, destruction and construction (or creation) are separable, contingent events. "Construction has to do with changing disorder into order, destruction with changing order into disorder" (Heider 2005: 11). Cheslatta T'En, as well as the Kayapó, Parakanã Indian nations, and a vast number of Indigenous nations worldwide, construct place-meaning through recognition and honouring of ancestral laws and kinship with place. Yet in the sweeping motion of colonial expansion, modernity and the disavowal of place value and order, destruction comes swiftly and dominates place relations.

The ordered elements of life upheld in place, include the organisms, houses, places of worship, social systems, pathways and hunting tracks, while those symbolising the disordered include submerged sites, floating coffins, suffocating layers of sediment, dams, death and disappearance of species. "Destruction is easy and can be very swift, while construction often involves a long and difficult process" (Heider 2005: 11), hence it cannot be easily replicated in a new place, or in the light of great harm and destruction.

> One can easily make disorder out of ordered entities, one can wreck them even if one had no idea what makes them work. For destruction the energy can be undirected and haphazard, for building up, it has to be guided ... In a launching action, one produces a certain change in the proximal environment, which then goes on by itself and eventually produces the desired change in the distal environment.
>
> (Heider 1958: 106, cited in Heider 2005: 11)

Having written extensively on mining-induced displacement, Downing (2002: 3; 1996) declares that the problems of displacement (induced by both mining and infrastructure development) and resettlement pose major risks to societal sustainability. The loss of physical and non-physical assets that comes with large-scale development projects threatens so many already if not soon to be impoverished communities. There are degrees to which displaced community members will come to experience insecurity, yet all displaced persons face the risks of landlessness, joblessness, homelessness, marginalisation, increased morbidity, food insecurity, loss of access to common property and social disarticulation. The loss of this "social geometry" creates existential questions that pivot around a people's socio-temporal-spatial order (Downing 1996: 34–47). Adapted from Black's (1976) earlier work on social geometry, Downing's work presses us to think about place and spatial presences through the lens of shapes, sizes and relative positions of figures and presences in space, a principle that echoes a nested sensibility. These come together to form a multidimensional amalgam of social space (see Black 1976), one that can, in moments of harm, be characterised by disarray and distress for people. That there is an existing socio-temporal-spatial order is evident in each of the wounding cases thus far noted, and conveyed even more vividly in those accounts of place designification that tell of another form of destruction and designification, namely ethnic cleansing, not as an abandonment of place, but as the tearing of human kin from the very fabric of place life.

Ethnic cleansing enacts another extreme form of destabilisation for people and place. It is the systematic forced removal of people from place by a more powerful other, often another ethnic group, with the intent to make place ethnically homogenous. This is high-level designification, where the removal of place's sedimentary layers requires expulsion of the actual persons that stand to substantiate place through knowledge and a socio-temporal and spatial awareness. Where people might ask routinely of their cultural universe, who are we? And where are we? (see Downing 1996: 36), in the aftermath of ethnic cleanings, place responds to the absence of its kin and often the arrival of new invasive presences. If place is made up of multifarious agents and has a sentiency of its own, might it be also be left asking, who am I? And where are you? While the origins of ethnic cleansing may be lost to us, Davis (2000: 693) declares that it "qualifies the twentieth century as one of the darkest eras in human history".

Patterns of place life are radically destabilised as a result of ethnic cleansing. The cessation of economic and subsistence practices can lead to major physical shifts in place, including an accelerated growth in numbers of feral species (both floral and faunal, and those normally kept at bay by knowledgeable human intervention), the encroachment of sands, and the silting up of waterways and wells. In the absence of kinship to regulate human life in conjunction with place character, ethnically cleansed places will often commence a slow decline. Places once rich and life-sustaining can become threatening and or desolate. As pathways become overgrown, seasonal rhythms disrupted and ancestors wait for a kin that will never return, the sentient being that is place is left to atrophy. The fascination with abandoned places and 'places reclaimed by nature', whether 'ghost towns' or places left scared by war and disaster, is evidenced by the documentation of 'forbidden and forgotten places' by popular and artistic culture. It is canvassed across a body of place literature concerned with decay, the spectral, lightless and gloomy landscapes, loss and healing (Cox and Holmes 2000; De Certeau et al. 1998; Edensor 2005, 2013; Gesler 2003; Jonker and Till 2009; Mayerfeld Bell 1997; Trigg 2006, 2009; Wylie 2007).

Recounting the vast number and horrific details of these expulsions is beyond the scope of what I can attend to here, and for the sake of this discussion the emphasis is on the realisation that the forced removal of people from place causes a form of 'root shock' for place. The impact of ethnic cleansing on place can be understood through reflection on proxemics and how human bodies interact with place. Proxemics introduces an intimacy to place relations, emphasising the study of non-verbal communication. Categories that resonate here are those of the haptic (touch), kinesic (movement) and chronemic (time). The emphasis is on the embedding and enmeshing of human life in the place world as a presence that both feels and is felt. The relational logic espoused throughout this book renders this a deeply communicative event, in that people feel the nature of place through their embodiment and place in turn feels the presence of human life. The weight and character of the haptic, the rhythm and impact of the kinesic and the way time is structured in place and communicated by place give meaning to spatial behaviour and this shapes and reinforces the character of place.

Through non-verbal communication, the proxemics of people in place is evidenced by the ways in which place is modified by human presence (through the building of structures, earthworks, or carrying out of farming, hunting and gathering practices). These form part of a non-verbal communication between people and place in terms of what place will allow, will support or reject. Not all modifications can be lasting and those that mesh best with the order of place will be the most sustaining. It is found in the way that place responds to human action, through such events as fire regimes and the vegetal politics of what will grow or disappear in a region, or through ancestors facilitating hunting and gathering across homelands without impediment. So too is human movement affected by the non-verbal communications of identifiable spatial limits of a region associated with a particular ethnic group. As with the olive groves of Palestinian villages, place is using non-verbal communication to unrest the movement of Israelis and to naturalise the presence of Palestinians.

Place is also capable of reflecting a chronemic tendency in which time passing communicates relationships with human presences and kinship. Seasonal shifts that are read for indications of animal breeding seasons, the alterations in mood that might come with the passing of day into night, the relationship between ceremony, light and dark and the synching of festival life with climatic adjustments (again seasonal shifts) are undoubtedly all modes of communicating a relationship between a certain people and certain place. Where places are ethnically distinguished through long histories of occupation and ancestral influences that have been pivotal to their making, the taking of people away from this nexus is an event that interrupts the proxemics of place. For if proxemics denotes spatial behaviour, and giving of structure and meaning to space (Hall 1966: 95), then the absence of the human to maintain this spatial meaning, ruptures the nature and order of place itself.

Despite evidence for a deeply relational encounter, and the likelihood of relational suffering, very little has been written on the matter of what happens to place once its people have been expelled. The land is rid of its people, sometimes even so violently as to involve the removal of all physical vestiges of the ethnic group through the destruction of earthworks, monuments, cemeteries, houses and religious centres. In fact, it is impossible to conceive of ethnic cleansing or genocide without a territorial dimension (Shaw 2007: 61). History, and in particular the Holocaust, has shown that territorial expulsion is often the precursor to complete or attempted physical extermination (Shaw 2007: 62). Lands will be burnt, crops destroyed, animals killed (both Indigenous and non-Indigenous, domesticated and wild), yet the bleak scape that remains is rarely examined for its own loss. Urbicide has received some attention, as the destruction of urban centres, the demolition of buildings, and fraying of urban fabric as a key element of war torn cities. It concerns the "killing, slaughter or slaying of that which is subsumed under the term 'urban'", also referred to as "rubbleization" and "saturation bombing" (Coward 2006: 427; 2004; Shaw 2007: 75). Nazi attacks, the Khmer Rouge's emptying of Phnom Penh in 1975 and the Serbian nationalists' campaigns against Sarajevo and other cosmopolitan centres in Bosnia-Herzegovina are illustrative of a distinctive violence targeted at urban and built places (Shaw 2007: 75; see also Halilovich 2011, 2013, for a powerful account of ethnic cleansing in Bosnia).

The emphasis on urban decay and threatening renewal in the literature on place harm eclipses an equivalent body of literature on place harm beyond the city limits. This triggers reflection on the fact that cities have historically been granted considerable agency in the place world. In popular writing and film, the city is presented to us as tamed space, yet a space that exerts its agency and influence over human life; it is seething, cruel, fast and capable of letting one in or keeping one out (Lefebvre 1996; Sennet 2000; Simmel [1903] 2002). It is animated through identity inscription, and may be regarded a noble widow (Minca 2009) or grand dame. Anthropomorphism plays a huge part in the inscription of meaning, with cities seen to have a wide range of body parts, including hearts, arteries, lungs, spines, while being capable of body functions, such as eating up and excreting waste (Solesbury 2013: 7). A particular form of political violence can be achieved in cities, where the elite alongside the impoverished may be affected by violence and destruction (Coward 2006). As urban space is more often defined by a shared spatiality, its destruction can lead to the obliteration of both mundane and sacred space, along with heterogeneity. Coward's writing (2006: 420) reminds us that by some logic of sense-making, the destruction of human life is one thing, but the destruction of the built environment another.

Urbicide and the concern for its effects reverses the anthropocentrism that is elsewhere applied to accounts of human suffering, ethnic violence and genocide. The reversal is found in the fact that human life is no longer afforded the greater value, nor is it given a higher ground in the experience of harm. Coward (2006: 422) promotes this through a discussion of urbicide as a distinct form of political violence, not dismissing the human toll, but bringing it to an equal standing with what the built environment undergoes when destroyed. He cites Croatian writer Slavenka Drakulic's *Mostar Bridge Elegy*, where the author grapples with the distinction between loss of human life and loss of a bridge. "The bridge was built to outlive us ... it transcended our individual destiny. A dead woman is one of us – but the bridge is all of us" (cited in Coward 2006: 420). Perhaps this is why urbicide more frequently arises in the literature on place harm and why Coward's (2006: 422) emphasis on the experience of place as one of receipt of political violence is of great interest here.

If the same principle is applied to all contexts and places in which human life is culturally and physically wounded, then the discussion of harm and healing is radically enhanced. What of the homogenous spaces in which the marginalised struggle to exist, where the built environment is not distinctive, but the ancestral mapping that inscribes it carries the greatest weight. What of the ethnic cleansing and destruction of these places? Is it that the impact of wounding and harm cannot so visibly be accounted for, or is it that the beauty of place, and the magnitude of its loss is overlooked because its qualities are maintained and safeguarded by a communing between non-human co-presences and multifarious agents, and that which only the culturally attuned are capable of seeing and hearing.

Chapter overview

This chapter was set the challenge of distinguishing humans as a troubling species of harm in a world marked by violence in place and the greater effects of cultural wounding. Contemporary and also historical instances of violence and harm have been canvassed for evidence of what motivates and inspires people to act against place and its people. The most obvious starting point for this diagnosis has been to assess the physical destruction and designification of place for human gain. It is contended through illustrative accounts, including the lived experiences of Cheslatta T'En, that what we see at work are axiological crises and simultaneous failure to care about the well-being of people and place. What distinguishes human action as the catalyst for violence and wounding is the intent to harm place and its human kin, and a disavowal of place law and existing order.

As processes of wounding, destruction and designification aim for the same outcome: removal and/or denial of place order. Relationships of affinity are challenged as places are harmed and imprinted with new identities as enforced by violent others. This is seen to be the case with aerial bombings in war times, and land theft and alienation as a signature of colonial frontier expansion. Historic and recent accounts attest to the occupation of territories, and ethnic cleansing of villages as violence directed at place, with the intent to harm people. Violence, in the service of cultural wounding, can erode the pathways that make possible relationships between people and place. Done in ways that attack the physical quality of place, these challenge an ethnic or cultural group's right to exist in place, their capacity to recognise and instate a relationship with place, thus tracing important connections to home. So too they undermine the nature of place itself and its ability to prevail in an identifiable form. If human kin struggle to enact relationships with place, then place struggles to maintain its distinctiveness. No longer does it have a witness to its distinct character and agency. This struggle is traceable to the constitutive rearrangement of place's character and physicality, which might occur in times of ethnic conflict and violence. The root shock becomes so profound that the fear is place may never be the same again.

Note

1 Windsor and McVey (2005: 155) report that the "non-natives who were displaced by the flooding were paid an average of $1544 per hectare of land and were provided with an average of two years' notice of the reservoir flooding". Meanwhile, the Cheslatta T'En were paid "an average of $77.22 per hectare for their land" (Windsor and McVey 2005: 155). In 1990, "the Cheslatta filed a $122 million court case against the Canadian government, and settled out of court for slightly less than $7 million" (Windsor and McVey 2005: 154). In 1998, members of the Cheslatta Carrier Nation filed a Statement of Claim with the Supreme Court of British Columbia disputing all agreements and licences issued to Alcan by the federal and provincial governments. The appeal was dismissed in Alcan's favour with no further legal actions in effect (Cheslatta Carrier Nation v. British Columbia 2000). On 30 January 2012, Rio Tinto Alcan finalised the transfer of approximately 11,000 acres of land to the Cheslatta Carrier Nation. The Cheslatta received the land as a form of freehold ownership with an estimated value of $1.2 million (BC Local News 2012).

References

Abufarha, N. 2008. Land of Symbols: Cactus, Poppies, Orange and Olive Trees in Palestine. *Identities* Vol. 15(3): 343–368.

Alexander, D. 1999. *Natural Disasters*. London: Kluwer.

Amazon Watch, nd. *Brazil's Belo Monte Dam: Sacrificing the Amazon and its Peoples for Dirty Energy*. Available at http://amazonwatch.org/work/belo-monte-dam. Accessed 12 August 2015.

Amir, S. and Rechtman, O. 2006. The Development of Forest Policy in Israel in the 20th Century: Implications for the Future. *Forest Policy and Economics* Vol. 8: 35–51.

Anderson, D. 2004. Reindeer, Caribou and 'Fairy Stories' of State Power, in *Cultivating Arctic Landscapes: Knowing and Managing Animals in the Circumpolar North*. Edited by D. Anderson and M. Nuttall, pp. 1–16. NewYork: Berghahn.

Bardenstein, C. 1999. Trees, Forests and the Shaping of Palestinian and Collective Memory, in *Acts of Memory: Cultural Recall and the Present*. Edited by M. Bal, J. Crewe and L. Spitzer, pp. 148–170. Hanover, NH and London: University Press of New England.

Bartolome, L.J., de Wet, C., Mander, H. and Nagaraj, V.K. 2000. *Displacement, Resettlement, Rehabilitation, Reparation and Development*. Working paper of the Working Commission on Dams. Available at http://siteresources.worldbank.org/INTINVRES/Resources/DisplaceResettleRehabilitationReparationDevFinal13main.pdf. Accessed 29 June 2015.

Barton, A.H. 1969. *Communities in Disaster: A Sociological Analysis of Collective Stress Situations*. Garden City, New York: Doubleday and Company.

Baumwoll, J. 2008. *The Value of Indigenous Knowledge for Disaster Risk Reduction: A Unique Assessment Tool for Reducing Community Vulnerability to Natural Disasters*. MA Thesis, Webster University, Missouri.

BC Local News. 2012. *Rio Tinto Alcan Returns Cheslatta Carrier Nation's Traditional Territory*, 8 February 2012. Available online www.bclocalnews.com/news/138880064.html. Accessed 13 August 2015.

Bebbington, A. (ed.) 2012. *Social Conflict, Economic Development and the Extractive Industry: Evidence from South America*. Abingdon, Oxon: Routledge.

Beck, J. 2010. *Three Towns: A History of Kitimat*. Victoria, British Columbia: Trafford Publishing.

Bidney, D. 1996. *Theoretical Anthropology*. New Brunswick, New Jersey: Transaction Publishers.

Black, D. 1976. *The Behavior of Law*. Bingley: Emerald Group Publishing Limited.

Blaikie, P., Cannon, T., Davis, I. and Wisner, B. 1994. *At Risk: Natural Hazards, People's Vulnerability, and Disasters*. London: Routledge.

Borup J.H., Gallego, D.T., and Hefferman, P.G. 1979. Relocation and its Effect on Mortality. *The Gerontologist* Vol. 19: 135–140.

Bradley, J. 1997. *Li-Anthawirriyarra, People of the Sea: Yanyuwa Relations with their Maritime Environment*. PhD Dissertation, Northern Territory University, Australia.

Bradley, J. 2000. Country of Our Spirit. Yanyuwa Land and Sea Scapes. *SAQ South Atlantic Quarterly* Vol. 98(4): 801–816.

Bradley, J. 2001, Landscapes of the Mind, Landscapes of the Spirit: Negotiating a Sentient Landscape, in *Working on Country: Indigenous Environmental Management of Australia's Lands and Coastal Regions*. Edited by R. Baker, J. Davies and E. Young, pp. 295–304. Melbourne, Victoria: Oxford University Press.

Bradley, J. 2010. *Singing Saltwater Country: Journey to the Songlines of Carpentaria*. Crows Nest, New South Wales: Allen and Unwin.

Bradley, J., and Kearney, A. 2011. 'He Painted the Law' – William Westall, 'Stone Monuments' and Remembrance of Things Past in the Sir Edward Pellew Islands. *Journal of Material Culture* Vol. 16(1): 25–45.

Braverman, I. 2008. The Tree is the Enemy Soldier: A Sociolegal Making of War Landscapes in the Occupied West Bank. *Law and Society Review* Vol. 42(3): 449–482.

Bunge, W. 1973. The Geography of Human Survival. *Annals of the Association of American Geographers* Vol. 63(3): 275–295.

CBC News. 2015a. Human Remains Wash Ashore in Flood as Anger Mounts over River Management, 28 May 2015. Available at www.cbc.ca/news/human-remains-wash-ashore-in-flood-as-anger-mounts-over-river-management-1.3091523. Accessed 10 August 2015.

CBC News. 2015b. Human Remains Wash Ashore in Nechako River Flooding, Rio Tinto Alcan Blamed, 28 May 2015. Available at www.cbc.ca/news/canada/british-columbia/human-remains-wash-ashore-in-nechako-river-flooding-rio-tinto-alcan-blamed-1.3089786. Accessed 10 August 2015.

Cherry, K. (ed.) 2009. *Lifespan Perspectives on Natural Disasters: Coping with Katrina, Rita, and Other Storms.* London: Springer.

Cheslatta Carrier Nation v. British Columbia. 2000. British Columbia Court of Appeal document. Available at www.ceaa.gc.ca/050/documents_staticpost/cearref_37519/45281/t12.pdf. Accessed 10 August 2015.

Christensen, B. 1995. *Too Good to Be True: Alcan's Kemano Completion Project.* Vancouver, BC: Talon Books.

Cohen, S.E. 1993. *The Politics of Planting: Israeli–Palestinian Competition for Control of Land in the Jerusalem Periphery.* Chicago and London: University of Chicago Press.

Colchester, M. 2000. *Sharing Power: Dams, Indigenous Peoples and Ethnic Minorities.* Working paper of the World Commission on Dams. Available at www.forestpeoples. org/sites/fpp/files/publication/2010/08/damsipsethnicminoritiesnov00eng.pdf. Accessed 29 June 2015.

Coward, M. 2004. Urbicide in Bosnia, in *Cities, War and Terrorism.* Edited by S. Graham, pp. 154–171. Oxford: Blackwell.

Coward, M. 2006. Against Anthropocentrism: The Destruction of the Built Environment as a Distinct Form of Political Violence. *Review of International Studies* Vol. 32: 419–437.

Cox, H. and Holmes, C. 2000. Loss, Healing and the Power of Place. *Human Studies* Vol. 23: 63–78.

Cunningham, S. 2014. *Warning: The Story of Cyclone Tracy.* Melbourne: Text Publishing.

Davis, D. 2000. Editorial. Confronting Ethnic Cleansing in the Twenty-First Century. *Journal of Church and State* Vol. 42(4): 693–701.

De Certeau, M., Girard, L. and Mayol, P. 1998. *The Practice of Everyday Life: Volume 2 – Living and Cooking.* Minneapolis, MN: University of Minnesota Press.

Downing, T. 1996. Mitigating Social Impoverishment when People are Involuntarily Displaced, in *Understanding Impoverishment: The Consequences of Development-Induced Displacement.* Edited by C. McDowell, pp. 33–48. Providence: Berghahn Books.

Downing, T. 2002. *Avoiding New Poverty: Mining-Induced Displacement and Resettlement.* London: International Institute for Environment and Development.

Dyson, M. E. 2006. *Come Hell or High Water: Hurricane Katrina and the Color of Disaster.* New York: Basic Civitas Books.

Edensor, T. 2005. *Industrial Ruins: Space, Aesthetics and Materiality.* Oxford: Berg.

Edensor, T. 2013. Reconnecting with Darkness: Gloomy Landscapes, Lightless Places. *Social and Cultural Geography* Vol. 14(4): 446–465.

Environment Canada. 2004. *Threats to Water Availability in Canada*. Ottawa: Minister of Public Works and Government Canada.

Erikson, K. 1976. *Everything in its Path: Destruction of Community in the Buffalo Creek Flood*. New York, London: Simon and Schuster Paperbacks.

Erikson, K. 1994. *A New Species of Trouble: Explorations in Disaster, Trauma and Community*. New York: W.W. Norton and Co.

Erikson, K. 2011. The Day the World Turned Red: A Report on the People of Utrik. *Yale Review* Vol. 99(1): 27–47.

Eyerman, R. 2015. *Is this America? Katrina as Cultural Trauma*. Austin, Texas: University of Texas Press.

Falah, G. 1996. The 1948 Israeli–Palestinian War and its Aftermath: The Transformation and De-Signification of Palestine's Cultural Landscape. *Annals of the Association of American Geographers* Vol. 86(2): 256–285.

Fearnside, P. 2006. Dams in the Amazon: Belo Monte and Brazil's Hydroeclectric Development of the Xingu River Basin. *Environmental Management* Vol. 38(1): 16–27.

Fienup-Riordan, A. 1994. *Boundaries and Passages: Rule and Ritual in Yup'ik Eskimo Oral Tradition*. Norman: University of Oklahoma Press.

Fisher, W.F. 1999. Going Under: Indigenous Peoples and the Struggle Against Large Dams. *Cultural Survival Quarterly* Vol. 23: 29–32.

Fullilove, M. 2013. *Urban Alchemy: Restoring Joy in America's Sorted-Out Cities*. New Village Press.

Gedicks, A. 2001. *Resource Rebels: Native Challenges to Mining and Oil Corporations*. Cambridge: South End Press.

Gesler, W. 2003. *Healing Places*. Lanham, Maryland: Rowman and Littlefield Publishers.

Good, B.J. 1996. Mental Health Consequences of Displacement and Resettlement. *Economic and Political Weekly*, 15 June, 1504–1508.

Gray, A. 1996. Indigenous Resistance to Involuntary Relocation, in *Understanding Impoverishment: The Consequences of Development-Induced Displacement*. Edited by C. McDowell, pp. 99–122. Providence: Berghahn Books.

Green D. and Raygorodetsky, G. 2010. Indigenous Knowledge of a Changing Climate. *Climatic Change* Vol. 100(2): 239–242.

Green, D., Billy, J. and Tapim, A. 2010. Indigenous Australians' Knowledge of Weather and Climate. *Climatic Change* Vol. 100(2): 337–354.

Halilovich, H. 2011. Beyond the Sadness: Memories and Homecomings Among Survivors of 'Ethnic Cleansing' in a Bosnian Village. *Memory Studies* Vol. 4(1): 42–52.

Halilovich, H. 2013. *Places of Pain: Forced Displacement, Popular Memory and Trans-local Identities in Bosnian War-Torn Communities*. New York and Oxford: Berghahn Books.

Hall, E. 1966. *The Hidden Dimension*. Garden City, New York: Doubleday.

Hall, A. and Branford, S. 2012. Development, Dams and Dilma: The Saga of Belo Monte. *Critical Sociology* 38(6): 851–862.

Hannigan, J. 2012. *Disasters Without Borders: The International Politics of Natural Disasters*. Cambridge: Polity Press.

Heider, F. 1958. *The Psychology of Interpersonal Relations*. Hillside, New Jersey: Lawrence Erlbaum Associates.

Heider, F. 2005. Violence and Ecology. *Peace and Conflict: Journal of Peace Psychology, Special Issue: Military Ethics and Peace Psychology: A Dialogue* Vol. 11(1): 9–15.

Hewitt, K. 1983. Place Annihilation: Area Bombing and the Fate of Urban Places. *Annals of the Association of American Geographers* Vol. 73(2): 257–284.

Jonker, J. and Till, K. 2009. Mapping and Excavating Spectral Traces in Post-Apartheid Cape Town. *Memory Studies* Vol. 2(3): 303–335.

Kearney, A. 2007. Place Spirit and Intangible Cultural Heritage in a Contested Land, in *Other Contact Zones: New Talents 21C*. Edited by J. Ensor, I. Polak and P. Van Der Merwe, pp. 120–142. Perth, Western Australia: Curtin University.

Kearney, A. 2009. Homeland Emotion: An Emotional Geography of Heritage and Homeland. *International Journal of Heritage Studies* Vol. 15(2–3): 209–222.

Kearney, A. and Bradley, J. 2009. Too Strong to Ever Not Be There: Place Names and Emotional Geographies. *Journal of Social and Cultural Geography* Vol. 10(1): 77–94.

Khalidi, W. and Elmusa, S.S. 1992. *All That Remains: The Palestinian Villages Occupied and Depopulated by Israel in 1948*. Washington, DC: Institute for Palestine Studies.

Lefebvre, H. 1996. *Writings on Cities*. Translated by E. Kofman and E. Lebas. Oxford: Basil Blackwell.

Liston, G. and La Canna, X. 2014. Cyclone Tracy: Aboriginal People Foresaw 'Bad Event' in Darwin. *ABC News*, 25 December 2014. Available at www.abc.net.au/news/2014-12-25/cyclone-tracy-warning-to-aboriginal-people-to-leave-darwin/5987974. Accessed 11 November 2015.

Long, J. 2005. *Emplanting Israel: Jewish National Fund Forestry and the Naturalisation of Zionism*. MA Thesis, University of British Columbia.

Mayerfeld Bell, M. 1997. The Ghosts of Place. *Theory and Society* Vol. 26: 813–836.

McCully, P. 1996. *Silenced Rivers: The Ecology and Politics of Large Dams*. London: Zed Books.

Mercer, J., Kelman, I., Transis, L. and Suchet-Pearson, S. 2009. Framework for Integrating Indigenous and Scientific Knowledge for Disaster Risk Reduction. *Disasters* Vol. 34(1): 214–239.

Mignolo, W. 2007. Introduction: Coloniality of Power and De-Colonial Thinking. *Cultural Studies* Vol. 21(2–3): 155–167.

Minca, C. 2009. Trieste Nazione and its Geographies of Absence. *Social and Cultural Geography* Vol. 10(3): 257–277.

Nicosia, F. 2009. Ecology, Embodiment and Aesthetics of Death in Hurricane Katrina. *Mortality* Vol. 14(1): 1–18.

Omeje , K. (ed.). 2008. *Extractive Economies and Conflicts in the Global South: Multi-Regional Perspectives on Rentier Politics*. Aldershot, Hampshire: Ashgate.

Orlove, B., Roncoli, C., Kabugo, M. and Majugu, A. 2010. Indigenous Climate Change Knowledge in Southern Uganda: The Multiple Components of a Dynamic Regional System. *Climatic Change* Vol. 100(2): 243–265.

Paine, R. 1982. *Dam a River, Damn a People? Saami (Lapp) Livelihood and the Alta/Kautokeino Hydro-Electric Project and the Norwegian Parliament*. Volume 45 of the Indigenous Working Group for Indigenous Affairs. Copenhagen: International Work Group for Indigenous Affairs.

Parmenter, B.M. 1994. *Giving Voice to Stones: Place and Identity in Palestinian Literature*. Austin, Texas: University of Texas Press.

Patridge, W.L. 1989. Involuntary Resettlement in Development Projects. *Journal of Refugee Studies* Vol. 2: 373–384.

Peacock, G., Hearn Marrow, B. and Gladwin, H. 1997. *Hurricane Andrew: Ethnicity, Gender and the Sociology of Disasters*. Abingdon, Oxon: Routledge.

Pelling. M. 2003. *The Vulnerability of Cities: Natural Disasters and Social Resilience*. London: Earthscan.

Petras, J. 2013. Brazil: Extractive Capitalism and the Great Leap Backward. *World Review of Political Economy*, 4(4): 469–483.

Pontes Júnior, F. and Beltrão, J.F. 2004. Xingu, Barragens e Nações Indigenas. *Núcleo de Altos Estudos Amazônicos (NAEA)*, p. 28. Belém: Universidade Federal do Pará.

Porteous, J.D. 1990. *Landscapes of the Mind: Worlds of Sense and Metaphor*. University of Toronto Press: Toronto.

Porter, L. 2015. Syria War: Drone Footage Shows Destruction of Aleppo's Heritage Sites. *Telegraph*. Available at www.telegraph.co.uk/travel/destinations/middleeast/syria/11468349/Syria-war-Aleppos-heritage-sites-destroyed.html. Accessed 11 November 2015.

Ramos, A.R. 1996. *Indigenism: Ethnic Politics in Brazil*. Madison, Wisconsin: University of Wisconsin Press.

Scambary, B. 2013. *My Country, Mine Country: Indigenous People, Mining and Development: Indigenous People, Mining, and Development Contestation in Remote Australia*. Canberra: Australian National University E Press.

Schnell, I. and Mishal, S. 2008. Place as a Source of Identity in Colonizing Societies: Israeli Settlements in Gaza. *The Geographical Review* Vol. 98(2): 242–259.

Sennett, R. 2000. Cities without Care or Connection. The New Statesman Essay. Available at www.newstatesman.com/node/137800n. Accessed 14 August 2015.

Shaw, M. 2007 *What is Genocide?* Cambridge UK: Polity Press.

Shaw, R., Sharma, A. and Takeuchi, Y. 2009. *Indigenous Knowledge and Disaster Risk Reduction: From practice to policy*. New York: Nova Science Publishers.

Shkilnyk, A. 1985. *A Poison Stronger than Love: The Destruction of an Ojibwa Community*. New Haven: Yale University Press.

Sibley, D. 1988. Survey 13: Purification of Space. *Environment and Planning D: Society and Space* Vol. 6(4): 409–421.

Simmel, G. [1903] 2002. The Metropolis and Mental Life, in *The Blackwell City Reader*. Edited by G. Bridge and S. Watson, pp. 11–19. Oxford and Malden, Massachusetts: Wiley-Blackwell.

Solesbury, W. 2013. *World Cities, City Worlds: Explorations With Metaphors, Icons And Perspectives*. Leicestershire: Matador.

Trigg, D. 2006. *The Aesthetics of Decay: Nothingness, Nostalgia and the Absence of Reason*. New York: Peter Lang.

Trigg, D. 2009. The Place of Trauma: Memory, Hauntings and the Temporality of Ruins. *Memory Studies* Vol. 2(1): 87–101.

Trigger, D. 2008. Indigeneity, Ferality and what 'Belongs' in the Australian Bush: Aboriginal Responses to 'Introduced' Animals and Plants in a Settler-Descendant Society. *Journal of the Royal Anthropological Institute* Vol. 14: 628–646.

Windsor, J. and McVey, J. 2005. Annihilation of Both Place and Sense of Place: The Experience of the Cheslatta T'En Canadian First Nation witin the Context of Large-Scale Environmental Projects. *The Geographical Journal* Vol. 171(2): 146–165.

Wood, J. 2013. *Home to the Nechako: The River and the Land.* Victoria, BC: Heritage House Publishing.

Wylie, J. 2007. The Spectral Geographies of W.G. Sebald. *Cultural Geographies* Vol. 14: 171–188.

Zerubavel, Y. 1995. *Recovered Roots: Collective Memory and the Making of Israeli National Tradition.* Chicago and London: University of Chicago Press.

4 Social disorder

Toponymic erasure and the making of harmful places

Social disorder in place can be achieved without physical acts of violence. The frontier upon which this violence and its wounding effects march is an ideological one. While not necessarily divorced from attempts at physical destruction and designification, social disorder achieves chaos in the hearts and minds of those who identify with and often cling to place as a presence of importance. So too it causes the substantive rearrangement of place character. Such experiences and outcomes of violence are tracked through the culturally constructed visions and wounding ideologies that lay claim to determining what a place is or is not. Through disordering events, a place can be transformed from a home, and meaningful co-presence in life, to a non-place left blank for the destructive acts of forced removal, nuclear testing or open-cut mining. It can be renamed, partitioned, its story erased and its elements destroyed.[1] Culturally constructed visions of place can come to compete in the place world. In these instances, existing place knowledge and character, as it is intimately linked to and known by specific ethnic groups, is challenged and new determinations are made as to what place should be and how its constituent parts might be arranged in the form of where people and their cultures belong. The disordering of place is achieved through the disregard of existing place identity and failure to recognise place worth. This can extend to how non-human species are arranged, the ontological order of place and its axiological merit.

Places marred by social disorder are often found in interethnic contact zones, providing the context in which determinations are made as to the presence and future of ethnic groups. It is by determining the type of human presence or enforcing absence that place can be left without its human kin or is compelled to accommodate the interloper as the 'other'. This other human presence may or may not recognise kinship as part of the relational encounter with place. The denial of kinship between people and place, and the rewriting of relationships into place, can cause a substantive disordering or reordering. This can be as harmful as acts of physical destruction and designification, often with consequences that affect the long-term survival of a place. By focusing on social disorder in place, consideration is given, in the first part, to acts of renaming. Renaming and the erasure of toponymic distinctiveness are the hallmarks of wounding events such as colonial expansion, and while renaming can be received positively in some cases, it is often a deeply contentious act that remains highly disputed for generations to come.

Another aspect of social disorder that is considered throughout this chapter concerns the brutal transformation of place from a site of belonging to one of containment and suffering. This concerns the creation of harmful places. As illustrated through narratives of enslavement and slave transportation, ethnic cleansing, mass graves and contamination, places once considered part of home territories and everyday life are transformed into places of harm and often terror. The creation of harmful places within existing landscapes means that places of belonging can no longer be separated from those of deep violence and harm. Home thus begins to exist primarily in relation to its loss. The ethnic group is then pressed to either remember or forget the violence and disordering of place. This triggers confusion in the kinship between people and place and may lead to one of several outcomes, retreat and no return, enduring in place and confronting the disorder, accepting new meaning of place or striving to return place to its original character. This form of violence and its resulting disorder is an effective instrument for the delivery of cultural wounding and intergenerational hardship, by inducing emotional geographies that link places in confusing and troubling ways. From this, people must finds ways to survive and continue to instate relations with the place world.

The violence of renaming and the erasure of toponymic distinctiveness

Place naming and narration are powerful vehicles for promoting identification and locating people and place within networks of memory (Alderman 2008: 195). The stories that are told of place begin by invoking its name. That a name can be called in the service of a people's identity is clear, as with the naming of nations, territories, rivers, oceans, border zones and communities. A name can also be denied and erased, subsequently replaced with another. In the case of the latter, renaming is an instrument of ideological warfare. Naming and renaming can also be a conduit for challenging dominant ideologies about place, past and identity.

The specifically harmful forms of renaming and erasures of toponymic distinctiveness that infuse this discussion are one part of a political violence that contests the existing rights of another to the possession, control and use of place; each of which convey a depth of kinship with place. They are also forms of emotional violence because they are, by design and practice, assaults on the emotional geography of place. Emotional geography specifically relates to the human emotions that are affected by place (Davidson et al. 2007), and engages with the spatiality and temporality of emotions as they coalesce around and in certain places. Emotions are triggered by and in relation to place, primarily travelling along pathways of kinship. Through naming, a place is made distinctive and this distinctiveness often resonates to inform the kind of emotional geography that it contains or is capable of generating in people's lives. A name can tell a great deal about place, particularly for those whose lives are enmeshed with the history of events or ancestral presences that are responsible for the making and naming of place.

Those instances when renaming might be received positively often signal breaks from a problematic past or dark history. Renaming may also signal the arrival of new freedoms, independence and autonomy, events that are often well-received by a majority populace. Widespread place renaming on such terms was initiated by the Zimbabwean government in 1982, in the hope of removing vestiges of British and Rhodesian rule (Guyot and Seethal 2007; Magudu et al. 2010, 2014). Erasing traumatic episodes of external rule, occupation or dispossession by way of renaming has been a documented practice in India since 1947, and more recently it has occurred with Myanmar and also across a number of cities and geographic sites in the Congo, which have seen name changes post-independence.

In Australia, momentum for changes to place names have gathered around some places that carry references to a racist and violent colonial past. Examples include the 'E.S. "Nigger" Brown Stand', a football stand that had its name changed after a nine-year battle, which led to the United Nations Council for the Elimination of Racial Discrimination (CERD) agreeing that the name was offensive and signage that displayed it should be taken down (a counter ruling to that made by the Federal Court of Australia) (Hagan 2006: 32; 2007). The stand has since been torn down with no name given to the new construction. Another example, 'Mt Niggerhead', in Victoria, Australia, was renamed 'Mt Jaithmathangs' after one of the Indigenous language groups of the region (Creative Spirits nd.). Other places for which changes are still being sought as part of an overall recognition of past wrongs and the horrors of British colonisation include, 'Black Gin Creek Road', and 'Nigger Road' in Queensland, Australia (Creative Spirits nd.). So too 'Gins Leap', 'Coon Island' and 'Massacre Island' in New South Wales, Australia, are names that dredge up the atrocities of colonial violence and subsequent place disorder (Creative Spirits nd.).

While there can be no denying that reinstating place names can work in the service of healing agendas for ethnic minorities, in the context of this discussion, the emphasis is on the intent to harm both people and place by erasing place names. This is a way of disordering place without destroying its tangible qualities. The impact comes from the silencing of place's intangible heritage, as it is contained in a name and expressed through narrative. Once again this is often the signature of settler colonialism, colonial cartography and spatial inscription, where the process of renaming is regarded an exercise in 'logos', the making of order, knowledge and language, over a space that is perceived as "passive, unknowing and feminized" (Herman 1999: 76). The space may equally be perceived as threatening, unknowable and untamed, hence it is in need of constraint and control, as can be achieved by instruments of recording and documenting. Naming and narrative erasure are, in instances of place harm, vital to the 'bringing in' or domestication of place. Taking control of place by delimiting its new potentials is frequently a deliberate step, taken with little consultation or care for those who share kinship with place or any place identity that may already exist.

Carter (1988: 61) describes the process of removing toponymic distinctiveness as a process designed to "neutralize otherness", where land, waters and resources are the ultimate desire of the invading presence. Roberts (1993), who writes of the political economy of place names, reflects on the naming practices of Indigenous groups, colonisers, government mapmakers and settlers in the Brazilian state of Pará. He identifies the extent to which "names contribute subtly and forcefully to our relationship with our planet" (Roberts 1993: 160). Some names contain evidence of the ecologically devastating violence a culture is capable of. For example, Atchison (1990) writes of the expression 'the outback' as introduced through Australian colonial discourse to illustrate the inherent links this name has to ideologies of separation, and an ontology of harm. According to Atchison:

> The term outback came into being to describe the regions remote from the settled districts. Its adoption, along with other terms, many of which came from the Aborigines, accompanied a rejection of the beautiful generics of an intimate English countryside ... Outback, originally a slang term and now orthodox, was accepted rather than *interior* which implies a spiritual, intellectual and emotional acceptance of the moderately arid shield and platform deserts with their distinctive vegetation, boulders and gibbers. White Australians are still learning to see the interior as a rich and diverse range of beautiful and delicate ecosystems. Aboriginal, not European transfer names,[2] fittingly evoke the spirit of these landscapes.
>
> (Atchison 1990: 153–154)

On themes of separation, disconnection and colonial imaginings of place, Rose (2004) has also written. She expresses the separation and disconnect through a language of ethics. The making of a 'ground zero', a deeply wounded space, where life is killed and potentially cannot return, is the essential ingredient for land alienation and subsequent destruction under the banner of colonisation. It is only from this ground zero that a new place order can be attempted. Yet this landscape, etched with scars does not disappear, and the moral failure of harm-doing is held to remain in place. This unsettling presence is what infuses settler colonialism and the enduring nature of unreconciled pasts in Australia. Its remedy is found, according to Rose (2004) in an ethics for decolonisation. For the settler in the colony, Rose (2004: 6) writes, "we cannot help knowing that we are here through dispossession and death". Place names are evidence of this, and the social disorder they trigger in place is aligned with the ethics of settler colonialism. They act as devices for writing upon the lands and waters of Indigenous homelands, attempting to silence if not kill an existing order of place. Rose (2004: 7) describes the action of 'doubling up' as a form of violence and resulting wounding that is symptomatic of frontier settings. As a form of "double violence", the doubling up of which Rose (2004: 7) writes is a "continuing act of wounding that not only kills parts of a living system but actually disables or kills the capacity of a living system to repair itself ... Doubling up is an amplification of death, such that death exceeds a balance with life and becomes a self-amplifying process itself" (Rose 2004: 7). This process is evidenced by the multiple methods of place harm that characterise settler colonialism; including place destruction and designification as achieved through physical erasure of place elements, forced removals and genocide, and the social disordering of place that comes with assaulting the toponymic distinctiveness of Indigenous places and homelands.

Patterns of naming are so reflective of the epistemology, ontology and axiology of which they are born that their very form says something of the affective relations of domination and subordination between social groups and the world (environment) more broadly (Roberts 1993: 160). Indigenous place names, as called into existence by ancestral beings, derived from their movements or inherent features of place and place-creating agents, reflect a fundamentally different epistemology, ontology and axiology to those generated and emplaced by colonial authorities and new settlers (Casey 1996; Ingold 2000; Hercus and Simpson 2002; Morphy 1995; Tamisari 2002). So too, as history and recent renaming suggests, the naming practices of a newly liberated ethnic group will differ starkly from those of their former oppressors. Roberts' (1993) work reflects the fundamental differences that emerge through naming and renaming, and highlights how these might disorder a place by writing into it a vast and competing number of narratives concerning place identity, worth and value. Undertaken at a time of intensive mineral exploration, deforestation and industry in the Amazon region, Roberts' (1993) research sought to document place names and patterns of toponymic distinctiveness. He recorded names associated with the Indigenous cultures and languages of the region, Catholicism, national hopes and aspirations, presidents and military personnel, natural features and other cities.

Competing for the majority of names were those of Indigenous origins and those with Catholic traces, and in some instances both were found to be simultaneously present in place. The toponymic distinctiveness of the region is an artefact of pre- and post-colonial imaginings. It reflects the events of ancestral creation, Portuguese invasion and occupation, attempts at justification for new arrival and aspiration for future place potential. There is a narrative in the naming of places throughout the region, which, according to Roberts (1993: 176), is accessed only by asking three key questions, "Who has the power to name places?" "What will their names represent?" and "How will we perceive the places because of their names?" Such questions are vital when we are reminded of the methodologies that guide this work; namely Indigenous pedagogies of place, emotional geography and nested ecology. It is such that the first of these methodologies complicates the way naming is understood and returns the focus to one of place agency and sentiency. In which case, the question is rephrased as who or what has the power to name place? It is by answering this, and in appreciating the role of place itself and non-human agents in the naming process, that a depth of understanding around what names represent and how these affect human perception and interaction is reached. Denial of place order and agency can be found in the stripping away of a place name, and also in the assumption that naming belongs solely to the realm of human cognition.

Most often configured as an act of human inscription, naming is understood as the laying down of cultural meaning onto space, as the world in which we live. Place names are taken to embed the politics of place and the roots of this action are then traced back to the workings of the human mind and cognition. Place names typically use a single word or series of words to distinguish one place from another and these toponyms are held to evoke powerful connotations (Alderman 2008: 195). In utilising an Indigenous epistemology of place and working to decolonise understandings of place, it would be remiss to leave unturned this normative understanding of place naming. It does offer a conceptual framework to understand some epistemological and ontological contexts in which place is brought into a relationship with human life, but is not comprehensive enough to fully expand on how it is that place names come to be. In which case, it is imperative to note that place names are not always externally granted and written by people into place. They are equally understood as inherently in and of place, and called by ancestral spirits long before the arrival of any people in place.

Certainly recorded histories attest to the common practice of people and political administrations laying claim to the determination of a place name, and through its utterance bringing about certain epistemological and ontological realities. But it is also widely held across Indigenous epistemologies that place names are the work of ancestral and spiritual beings and non-human agents. A place name can very easily pre-exist human life and stand as a marker of what always was and always will be. No attempt at renaming can undo the inherent place order that came with ancestral makings and naming. Casey (1996: 26) refers to some places as "ideolocal", meaning that in some cases a "place is more an event than a thing to be assimilated to known categories". In the case of Indigenous Australian homelands, names come about through a process of morphogenesis. As Tamisari explains:

> At certain stages of their journeys, Ancestral Beings metamorphosed, imprinted and externalised whole or part of their bodies into topographic features such as a stone, a hill, the trees along a river, clouds, lightning, animals and plants. Ochre quarries are said to be ancestral faeces, a yellow clay which turns red after cooking is their blood, and clouds were formed by the water which spouted from their mouths ... During their wanderings over the land, sea and sky, Ancestral Beings and the objects used by them in a certain manner, left a mark behind.
>
> (Tamisari 2002: 93)

"Landscape features thus shaped are not only transformations of ancestral body parts but the embodiment of particular actions which identify unique cosmogonic events" (Munn 1996: 457, cited in Tamisari 2002: 93). It is the uniqueness of these 'action features' that is "condensed in the toponyms and other proper names given by the ancestors in the act of shaping the land and bestowing tracts of country upon people. Thus names may be seen as localising events, manifesting actions and congealing movement" (Tamisari 2002: 93). There is a profundity to names that exists because of these events and they set the terms of reference for all kinship with place. So too they narrate place into a wider order of places and actors, human and non-human. Accounts of place naming such as this are not isolated to Indigenous Australian epistemologies but can be found throughout the literature on global Indigenous place naming practices (Basso 1988, 1996; Brattland and Nilsen 2011; Cajete 1999; Deur 1996; Hercus and Simpson 2002; Nash 1999; Salmón 2000).

In conducting ethnography with Indigenous families in northern Australia, I have been introduced to an expression that begins to tell more of the importance of place names and place order through naming. The term *wirrimalaru* is an encompassing and rich qualifier used by Yanyuwa to describe place, people, ceremony and ancestors. *Wirrimalaru* is used most often to reference places and elders 'of great and inherent power'. A place identified as *wirrimalaru* is one governed by extensive ancestral knowledge and deep cultural meaning. In turn, the actual name given to a place of this high standing is born in place, granted by the ancestral beings that created place. Names and naming do not belong to the realm of the human, but instead are part of the DNA of place, and exist without human intervention. Thus the utterance of a place name is akin to the place giving you its name, and having that name sit upon your tongue; such is the possibility for agency beyond the human realm. One place that is regarded *wirrimalaru* is Manankurra (see Kearney and Bradley 2009a, 2009b), located on the eastern boundary of Yanyuwa homelands in northern Australia. The naming of this place, and the human utterance of this name, is described as follows:

> While the place name may be used in regular conversations about homelands and old times, any direct and detailed conversation about Manankurra requires certain etiquette. People often begin by asking; 'is it alright if I call the name of that place?'–*yamulhu barra karnawandarrbala na-wini awara*? This is a question usually asked by someone who is deemed junior to the listeners. It gives the listeners the chance to say 'Yes, keep going' or sometimes to abruptly say, 'No, you do not know that place!'–*kurdardi, yinda manji barra ki-awaruwu*!
>
> (Kearney and Bradley 2009a: 86)

In the case of the latter, the presumption that underpins the outright calling of a place name is tantamount to saying that you acknowledge the powerful Law that resides in place and therefore apprehend the many levels of knowledge about that place as well. If one does not know a place then one is not positioned intimately enough or does not yet emotionally comprehend the value and importance of this place, its name and creator ancestors. There is an emphasis on knowing the order of place as it has been shaped and passed on by the ancestors. The right to name or call this place depends on a number of things, including kinship with place, familiarity with its Law, knowledge of place order and permission. To call a name without authority can result in the individual being seen to act with disrespect and to be out of order, in which case he or she does not know what that place contains and does not comprehend its Law, thus insulting the ancestors and the place.

For each instance where place names are in and of place, beyond human acts of inscription, there are certainly instances in which people actively craft and attribute names to place. Acknowledging this is to realise the presence of distinct cultural traditions, as reflected by different epistemologies and ontologies. Highlighting the range of possibilities when it comes to place names does, however, expand the discussion on how renaming might be an act of violence. However they come to be inscribed, names are an indication of documented power.[3] That the use of this power can be unsettling is evidenced by the literature on place naming, which identifies two primary concerns, first, the act of renaming as part of a process of claiming territory and subordinating Indigenous histories and, second, naming as a postcolonial process of recovering lost language and memory (see Alderman 2008: 196; Berg and Kearns 1996; Herman 1999; Nash 1999; Roberts 1993; Yeoh 1992, 1996). Renaming as part of the violence of social disorder in place can have generation-spanning emotional effects on those living within occupied territories, or for those who must come to terms with expulsion, land theft and toponymic erasure. Renaming to advance notions of ethnic belonging and renaming as part of a process of claiming territory are detailed in the account of violence that follows.

It is with the combined sensibility of an anthropologist and cultural geographer that this second account of place harm is presented. This narrative of place harm describes the instating of social disorder in place, through the combination of warfare and military intervention, partitioning, renaming, forced relocation, generational tensions and ongoing disputes over territories. This case involves the partitioning of a single nation, populated by different ethnic groups. The resulting two territories have taken on distinct names, the south recognised internationally, the north a self-declared state, denied international recognition. Some have described the north as imagined, as 'make-believe', yet its form of 'made-upness' figures as "real space", but space that has been deliberately crafted (Navaro-Yashin 2012). To modulate the disorder that exists in place courtesy of the partitioning and corralling of ethnic groups to the north and south, United Nations peacekeeping forces are present and enforce a buffer zone designed to keep ethnic groups from coming into contact and transgressing boundaries, both actual and imagined, of enforced separation. This place is the Republic of Cyprus, or Cyprus and the Turkish Republic of Northern Cyprus.

Disordering Cyprus: intercommunal conflict and the demilitarised 'Green Line'

How one names place in this instance depends on the political geography that is recognised and the emotional geography with which one might empathise. The history of conflict in this region is well documented and reflects a tangled heritage of claims to territory. In 2003, I was invited to be a guest at the Eastern Mediterranean University, in the Turkish Republic of Northern Cyprus. Before leaving Australia, I had found myself in deep discussion with two friends, both of Greek Cypriot descent. They had asked me why I was visiting this place and what I thought of the history of conflict and ethnic tension. In reality, both were affronted by my plans to visit the north and shared their views of a deep animosity that had caused the division of the island nation. As I listened to their views, forthcoming from young Greek Cypriots born and raised in Australia, I was struck by the legacy of conflict and the extent to which they had absorbed some effect of the partitioning into their lives a great distance away in Australia.

Elsewhere I have written on the long-term and cross-generational effects of cultural wounding, an insidious side-effect of ethnic violence metered out through ethnic cleansing, forced removals, genocide and stigmatisation (Kearney 2014). In hindsight, I see strong evidence of cross-generational cultural wounding in the protests of my Greek Cypriot friends. Somehow, in the years since actual armed conflict and at a great distance from the actual site of conflict, they have absorbed the effects of their ancestor's losses. Knowing something of their family backgrounds, it is clear they have been enculturated to care for their ancestral home, and have lived with the social memories and oral testimonies of family members who recall the 1974 conflict and its ongoing expressions. I have since returned to the Turkish Republic of Northern Cyprus on three occasions. On my first visit to the north I requested my passport not be stamped, and while access for foreigners has become unhindered since 2003, each time I have visited the north I have felt a deep discomfort at my presence in a place so marked by separation and so disordered by contest. Furthermore, the dissent of my Greek Cypriot friends has continued to ring in my ears.

Navaro-Yashin, who has carried out long term ethnography in Northern Cyprus, writes on the impact of intercommunal tension and the lasting effects of colonial dissolution in the 1950s, along with the 1974 partitioning of Cyprus in the aftermath of the invasion of the north of the island by the Turkish army (see Navaro-Yashin 2002, 2005, 2006, 2009, 2010, 2012). Her work is at once deeply reflexive and richly anthropological, and sheds light on the forms of social disorder that have come to the island as a result of conflict and partitioning. She writes of how those resident in the north live alongside the properties and objects of the "so-called enemy". Communing with the principles espoused in this book, Navaro-Yashin (2009: 1) writes of affective geography and proposes an anthropologically engaged theory of affect through an ethnographic reflection on spatial and material melancholia. She documents a powerful sense of non-human agency in place, as it is found in the "emotive energies" discharged by objects, properties and places. By extension, she is writing of the agency of place and the ways in which place disorder might reveal itself through communicative events and pathways such as emotional geographies. People's grappling in the north with the enduring presence of those displaced (from the north to the south), as present in the form of objects, places, villages, streets and place names, is a powerful articulation of disorder in place. It is as if the elemental parts of place do not sit well when arranged alongside another ethnic presence, or as though the means by which this presence has come to be in place cannot be reconciled and put into healthy accord.

A strategy of 'space purification' took place across Cyprus, and in particular in the north as the Turkish forces sought the removal of ethnic others. This removal did not, however, involve the removal of other place elements, such as the built environment and the region's ecology. These features were left 'in place' as they were required to house and accommodate the incoming Turkish Cypriot population, along with military personnel. Thus, in leaving behind trace elements, the possibility of ideological sediment and evidence of Greek Cypriot place order seeping up through the layer of Turkish Cypriot presence was and remains highly likely. The threat and reality of this is, in large part, what Navaro-Yashin (2009) records ethnographically.

In the north, and also the south, place communes with people, and in doing so has the capacity to convey something of what is right or wrong, not necessarily in terms of a morally charged register of good and bad, but eschew in the order of how place once existed. Bearing in mind Navaro-Yashin's commitment to ethnography in the north, her proclivity for affective geographies and the recognition of non-human agency in place, there is no better way to start this account of place disorder than with her own description of impending place harm. She sets the scene as follows:

> Consider an island space, not too distant from Western Europe, where communities that had coexisted for centuries, if with tension, have begun to assume distinctly separate national identities, entering armed conflict with one another. Picture this taking place late in the 1950s, at a time of colonial dissolution and the formation of a sovereign nation-state meant to represent all the communities on this island. Strife between the two dominant communities survives the declaration of independence from colonialism, and, soon, these communities begin to move into separate, ethnically defined enclaves in faction with one another. In 1963, members of the community that is the majority on the island commit significant atrocities against members of the minority. Imagine these two communities, now already defined as distinct 'political communities', further

divided from one another with the arrival of an external army, which invades the northern part of the island, declaring that it does so in the interest of the minority.

(Navaro-Yashin 2009: 1)

The 'Green Line' began its life in 1964, as a ceasefire line to constrain intercommunal conflict between Greek Cypriots and Turkish Cypriots on the Mediterranean island of Cyprus. It became an impassable divide between the northern and southern parts of the state in July 1974. The divide was physically instated as a result of the Turkish invasion of Cyprus, during which Turkey captured Cypriot territory. The Turkish invasion traced its roots back to earlier intercommunal violence between Greek Cypriots and Turkish Cypriots that began in the 1950s and peaked in 1963 and 1964 (Borowiec 2000). These conflicts had resulted in the displacement of a vast number of Turkish Cypriots and the end of their political representation in the Republic (Papadakis et al. 2006). By 1974, deeply contested rights to be in place had simmered for more than a decade, and resulted in renewed vigour to reinstate Turkish Cypriot influence over the island.

By way of a military invasion, Turkey sought to take territories, assert political rule and expel Greek Cypriots from the north. "With the arrival of Turkish troops on 20 July 1974 and their subsequent invasion of towns and villages in Northern Cyprus, thousands of Greek- and Turkish-Cypriots were turned into refugees, leaving behind their natal villages, homes, land, and belongings and moving to the side of the island designated for their separate habitation after the war" (Navaro-Yashin 2009: 3; see also Loizos 1975). "Turkish-Cypriots who happened to live in southern towns and villages escaped to the north, now under Turkish sovereignty, to protect themselves from Greek nationalist reprisals. Greek-Cypriots who lived in the north moved in great numbers to the south, experiencing and fleeing major brutalities during the war" (Navaro-Yashin 2009: 2). People found themselves adrift, having lost their homes, land, gardens, animals, personal belongings, communities and jobs. With the transfer of people from north to south and south to north, and amid the fleeing, kinship with place was ruptured and both distance and dismissal have since prevented meaningful return and the reinstating of these powerful relationships with place.

The Green Line established the present and persisting divide between the zones of the north and the south. It also set limits to the spatial identities of the Greek Cypriots and Turkish Cypriots. The Green Line is a demilitarised zone that runs for more than 180 kilometres between the two partitions of the island and is patrolled by the United Nations Peacekeeping Force. To the south, control is held by the Greek government of Cyprus, and to the north, the Turkish Republic of Northern Cyprus. The distinctions separate and isolate ethnic groups and make specific claims to territory. Having viewed the Green Line from Famagusta (Turkish name: Gazimağusa; Greek name: Αμμόχωστος) in the northeast of the island, and Nicosia in the south of the island, the view is a saddening one. There are signs of decay, bullet scarred buildings, barbed-wire fencing, sandbags and concrete wall segments, watch-towers, anti-tank ditches and minefields. So too there is a silence and sense of feverish abandonment, which has been followed by the slow creep of absence. Wildlife has taken over in parts, returning in the wake of human departure. It is as Bryant (2012: 336) writes, a "gash through the island" that one side seeks to eliminate and the other to solidify. Standing on the beach at Famagusta, one can look back toward Varosha, a city fenced off by Turkish forces at the time of conflict in 1974. Once a thriving tourist resort, Varosha is now a scene of military presence and decay; best described as an empty city full of aging infrastructure on the verge of collapse.

It is not uncommon to meet Greek Cypriots on this beach, returning to look out at Varosha and see what remains of their former home. In 2003, I stood on this beach and watched a Greek Cypriot man (who had crossed the Green Line for a day trip[4]), as he focused a telescope onto the crum-bling façade of this former resort town. I learned that he was there to check on his apartment, one he had left behind in the rush of expulsions in 1974. Located on the waterside of Varosha, the apartment, his daughter explained, was frozen in time, sitting there waiting for them to return. He lamented, both through frustration and sadness, an altogether natural response to the depressing decay of Varosha and the loss of his property and possessions. Varosha gave little back, instead it stood still as a mess of abandoned apart-ment blocks, empty beaches and decaying signage. Quiet roadways lead to empty buildings, businesses, car dealerships and supermarkets. There are only dusty remains left to speak of the former residents having moved through and holidayed in this place. At this point, on the beach at Famagusta it is possible to follow the barbed wire fencing that delimits the boundary of military con-trol into the ocean. It goes below the surface and eventually gives way to sand. It is altogether possible to swim out past these boundary limits, transgressing the military line, yet few appear to do so. The question is; what would a per-son be entering if they stepped across this boundary? Varosha is on the one hand an occupied territory, yet it is empty, it is neither home nor away. It is a place that lacks the active relational presence that might give it meaning. In many respects it is a wound, a scar on the landscape that appears, at present, to have no capacity to heal.

On both sides of the conflict it is well documented that parties perpetrated violence. Ethnic cleansing, massacres and rapes have all been documented as crimes committed in the name of war and in the pursuit of territory. Accounts of disappearances continue to plague both the north and the south, yet charges of ethnic cleansing are more often levelled at the admin-istrations of the north (Council of Europe 2001; Kovras 2012; Sant Cassia 2005). It is claimed that some 2,000 people, both Greek and Turkish Cypriot 'disappeared' as a result of political violence during the two major periods of conflict in 1963 and 1974. Sant Cassia (2005: 2) notes that this repre-sents a significant number for a population of 600,000 people, contending further that "few bodies have been recovered: most will probably not be". The significance of the disappeared cannot be understated in the context of this book, as their absence hints at the possibility of unmarked graves, mass graves and deep traumas resting beneath the surface of the land. Bryant (2012: 333) also writes that since the opening up of the partition in 2003 for traffic to and from the north/south, many who had long searched for loved ones, assuming them trapped in enemy territory, are now faced with the awful reality that they have disappeared.

Mass graves containing victims from both sides of the ethnic divide have been found in the north and south (Sant Cassia 2005). In 2007, a list of missing persons was compiled and the United Nations-backed International Commission on Missing Persons was established, which involves a bi-communal effort between the Greek Cypriot and Turkish Cypriot communities. The Commission has led the search for burial sites, organising the excavation, exhumation and identification of bodies. These revelations expose past violence and offer up a deeply confronting sediment that returns Cyprus to the traumas of ethnic violence at unpredictable intervals. That these events have emotional and psychological effects on people in place is undeniable and complicates the identity of place in the minds of both ethnic groups.

The right to possess place, to assert an ethnic identity in place and to see a self-evidentiary aspect of identity in place was frayed heavily by the 1974 conflict. In turn, "two ethnically defined communities have been separated from one another with borders patrolled by armies on either side, not allowed to visit the other part of the island, each other, or their natal villages and towns for three decades, with access across the border banned" (Navaro-Yashin 2009: 3). In the wake of the partitioning, many places have been renamed. Those in the north carry Turkish nomenclature, those in the south primarily Greek Cypriot place names. Some villages were left to sit in the buffer zone. Pyla, Deneia, Athienou and Troulloi are some of the places left in this "Dead Zone" or "In-Between Zone" (Papadakis 2000: 93). Pyla, the most notorious mixed village in Cyprus, is home to both Greek Cypriots and Turkish Cypriots. Here they live together, yet divides are expressed through everyday separations in public space and a liminality that has, since 1974, marked it as a place of uncertainty and ambiguity (Papadakis 2000: 96).

Papadakis (2000: 93) explains that "[i]t is controlled neither by the Greek Cypriot, nor the Turkish Cypriot authorities, but lies under the jurisdiction, or better the supervision, of the United Nations force stationed in Cyprus, which has received the mandate of preserving the status-quo until a political solution is found". It is a place regarded with some curiosity, if not caution, as its residents, on both sides of the ethnic divide, were once considered "morally and patriotically suspect", in light of their avoidance of serious conflict during the 1974 war. Few were dislocated, in large part due to early UN intervention and "conscious preventative action taken by its own residents" (Papadakis 2000: 94). Despite the avoidance of serious conflict, this place lingers in the "Dead Zone", providing a fertile breeding ground for suspicions

in times when ethnic tensions reignite. Liminality and the impending confusion over place identity in Pyla has an effect on place as drastic as processes of exclusion and alienation experienced either side of this in-between
place. Radiating out in both directions from the Green Line, one encounters
other forms of social rearrangement that are reflected in the presence of new
place names, retention of old place names, new residents in old houses, old
residents in new houses. In the north, it has been argued that residents have
refrained from investing in the land because of the uncertainty that came with
living in occupied territory and "Greek-owned land", and lived experience of
place as marred by confinement, entrapment and suffocation (Navaro-Yashin
2009: 12).

Writing on the experiences of those who fled north, namely the Turkish
Cypriot refugees, Navaro-Yashin (2009: 3) describes that they "were mostly
allocated Greek Cypriot houses, land, and belongings by the Turkish-Cypriot
administration in the north". The refugees found themselves using, inhabiting, employing and interacting with spaces and properties left behind by the
former community (Greek Cypriots) (2009: 3). Navaro-Yashin writes:

> When they appropriated Greek-Cypriot villages, land, and houses,
> Turkish Cypriots often recounted to me stories of how they found rotten
> food on set dinner tables, left behind by Greek-Cypriot evacuees at the
> sound of sirens when Turkish troops arrived in Northern Cyprus. Others
> gave accounts of suitcases full of personal belongings, which they found,
> thrown on the side of the road, by fleeing Greek-Cypriots unable to carry
> the weight in the moment of escape. Turkish-Cypriots appropriated these
> leftover objects and belongings. Having been stripped of their own things
> in the south as they were displaced to the north, some refugees had to
> wear garments left behind by the Greek-Cypriots, cleaning them in the
> rivers.
>
> (Navaro-Yashin 2009: 2)

Written with a gentle yet great aptitude are Navaro-Yashin's accounts of
'looting hunts' in which people visited Greek Cypriot villages and houses,
taking furniture, appliances, sheets and bedding. "All these objects were
picked from the houses of Greek-Cypriots, from amidst their personal spaces
and belongings, with family photos hanging on the walls, personal diaries in
draws and books with names inscribed in them" (Navaro-Yashin 2009: 2). The
effect of these relations with place, and the absent, yet present Greek Cypriot
other is, according to Navaro-Yashin (2009: 5) "melancholia". Melancholia
is not only drawn from the inner dialogue of living in an occupied territory
and the discomfort that comes with a lifelong awareness of constraint in
border crossing and access to other parts of the island, but also a feeling that
place "inflicts" itself upon the Turkish Cypriot residents. There is no doubt
that what Navaro-Yashin raises is the possibility of agency in place and the
response of place to its social disordering. She asks: "Is it Turkish Cypriots'

conflicted subjectivity that exudes an affect of melancholia in Northern Cyprus, or is it the rusty and derelict environment kept visibly unmaintained since the war that generates this feeling?" (Navaro-Yashin 2009: 6).

This place has been upturned and disordered in many ways and relations to place have continued to shift as relations between the north and south also change. Generations grow up in a partitioned world, yet one deeply inscribed with meanings and accounts of ideological sediment. After a nearly 30-year ban on crossings, the Turkish Cypriot administration significantly eased travel restrictions across the dividing line in April 2003, allowing Greek Cypriots to cross into the north and Turkish Cypriots into the south. This has brought with it a vivid realisation of what has been lost, and what remains at stake (Bryant 2012: 333). The muddled effects of social disorder in place linger as ongoing tensions between ethnic groups and unreconciled thefts and possessions that at times threaten to simmer over.

In Cyprus, as populations faced upheaval and displacement, place also lost constitutive parts of itself, as evidenced by renaming, abandonment and the establishment of no-go areas and buffer zones. In the case of the latter, place is literally stripped of prior identity and rendered a non-place. Such non-places populate geographies of 'nowhereness' and 'otherness'; and may trigger a crisis of identity for both the place and the people (see Arefi 2007). This is non-place configured somewhat differently to earlier framings by Webber (1964) and also Auge (1995). It is not the dystopia and emptiness of urbanity, nor is it transience that disregards the possibility of space being place. It is a statement of liminality combined with 'nothing-in-place' as an ideological and axiological preference. In 1974, Turkish military forces sought to neutralise place, strip it of existing identity and order in an attempt to instate new presence and order. Yet the passage of time has shown that this does not occur easily. Disordering events that saw the expulsion and forced relocation of persons to the south and north, as well as mass death and violence, have not been accepted into place as a new order. As an agent, place has been active in the rejection and acceptance of certain absences and presences, as reflected in the melancholia of living in occupied places, where disorder lingers and often prevents peace.

Social disorder, such as that caused by partitioning and renaming, can cause constitutive parts of place to be inscribed more heavily than others in social memory. This is the result of threat or loss, which triggers a hypersensitivity to place as part of an ethnic identity. Clinging to the memory of place and possessing its spirit through remembrance is frequently reported in instances where home is lost or alienated. The converse of this is the possibility of forgetting social memories associated with place, as people struggle to enact relationships that ensure remembering and enlivening of social memory. In this case, place character and identity may drop from consciousness and fall out of circulation as part of a fragmented or broken collective memory (Bakshi 2012). Partitioning, and renaming, when combined with the violence of expulsion and forced removal, can lead to a breakdown in transmittable knowledge of place, a condition that can be exacerbated by post-traumatic stress and external pressures that delimit the possibility of recalling place and enacting kinship. Such pressures may constrain opportunities to practice place knowledge, or the stigmatising of place knowledge and ontologies that allow people to enact kinship with place. The place world of a particular ethnic group can shrink in the midst of knowledge loss. If alienated from the places of their ancestors, many ethnic groups will struggle to maintain an intimate knowledge of who they are, with pressure added if there is a prevailing dominance of other authorities and new land tenure arrangements. This may conspire to form a type of generational forgetting, in which knowledge of the place world, and capacity to enact kinship with place is lost or significantly eroded.

On this matter of forgetting, Connerton (2008) identifies seven conditions that influence social and cultural life. Four of these resonate with forgetting as it might occur in the midst of social disorder, and trace back also to forgetting precipitated by destruction and designification. Forgetting of place character occurs when social disorder is linked to repressive erasure, new identity formation, forgetting as annulment, or prescriptive forgetting (Connerton 2008: 59). The first of these, repressive erasure, concerns the totalitarian and brutal erasure of memory structures and provocations. Achieved through destruction and removal of traces, renaming, destruction and designification are all signatures of repressive erasure. When place disorder is required to accommodate a new presence or the names of such new presences, then forgetting also becomes an essential part of new identity formation. A type of coercive forgetting is imposed on those who remember, while a form of instrumental forgetting may be deliberately adopted by those who wish to forget and for whom forgetting brings some benefit. In the case of the latter, forgetting becomes an act of service to those that wish to generate new meaning in place. The nexus between people and place is deliberately broken and resignified in the hope of it becoming an anchorage point for new identities. That these identities will always take hold, however, is no certainty.

In the context of place harm, Connerton's distinction of 'deliberate forgetting to move on' may occur as a strategy to silence and erase cultural and ethnic distinctiveness in place. This is closely aligned with forgetting as annulment, whereby the instating of new meaning in place is dependent upon forgetting what has previously existed, along with what may inherently be a part of place identity and meaning. Place agency and distinctiveness that exists becomes crushed beneath the weight of what is newly valued or devalued in light of the axiological crisis that supports place destructions and disordering (Connerton 2008: 64–65). As part of this project, there may emerge a contemptuous relationship in which place is indecipherable to people or where deliberate forgetting cannot be wholly achieved due to residual expressions of place-meaning, dissenting voices that will not give way to oppressive resignifications of place and a place order that continues to express itself even in the face of destruction and disordering acts. It may be that fauna, flora, seasons and elements pose a threat to the incoming group by simply replicating their patterns of order. These may be in stark contrast to the interloper/occupier's own sense of place, and lead to difficulty in successfully apprehending the character of place or flourishing in place. In reality, kinship between people and place is not easily found when presence is prefaced on the need to forget existing order. The fraught nature of rein-scriptive place-making is revealed by the stifling effects of living among the remnants of the so-called 'enemy' (see Navaro-Yashin 2009), and persisting melancholia, in parts of Cyprus.

Indeed, Navaro-Yashin's (2009) work highlights the difficulty involved in actually forgetting and the incompleteness of erasure designed to obliterate or overlay prior meaning in place. Gains are held to come only when one lets go of certain memories and knowledge of place, which means separat-ing place from its ideological sediment. Renaming can cause enough social disorder so as to begin this process, yet may not always be enough to silence the local empiricism and order that is place. Recognising this is vital to engag-ing with the sentient quality of place and its agency in times of crisis. On a spectrum from 'survival to death', place has the capacity to go either way as a result of social disordering. Human intervention, communing and the instat-ing of kinship are key determinants in what will happen and the extent to which disorder 'takes place' and strips it of all meaning. The role of place in accepting or rejecting human presence is something that must be considered. If a human presence is outside of the order of place, or conflicts with the ideological sediment that gives place its character, then the ill fit may mani-fest in complicated ways, from human depression and sickness to an inability to harvest food and find water. The agent that is place is not easily crushed beneath the weight of incoming influences and may assert its character to the detriment of newly arrived human presences.

It is in the liminality or indefinability of place as disordered or its becom-ing a void, that human loss is so powerfully recalled, and it is in light of a perceived nothingness in place that some of the greatest place harms are enacted. Social voids can develop in place as erasures of this kind take hold. Social voids are possible only when place is stripped of or denied agency and meaning, and is reduced to its spatial field. The void is the dismantled space in which meaning struggles to hold and find legitimacy. From this may come the experiences of disorder, whether as melancholia (Navaro-Yashin 2009), or as conveyed in the ethnographic accounts of Bakshi, (2012) as a "visceral sense of discomfort" in place. Social voids are defined by either their 'noth-ingness' as empty of meaning, or an indefinability whereby meaning is obfus-cated. This can be the experience for those who find themselves at home in a place no longer recognisable for the violence of disordering, or for the new-comer who cannot identify place character in ways that allows them to move successfully through place, thus demanding a form of kinship for successful navigation. The void reflects the magnitude of disorder in place and may be considered place's response to disordering events.

Kinship can act as the inhibitor of this liminality, indefinability and nothingness, as it is through recognised agency and co-constituting presence that people may find pathways to connect with place and affirm its character and order. This can prevail in the most shocking of circumstances, where homes, villages and communities have been transformed into sites of violence and suffering and may even prevail long after place has been lost and people expelled. The asserting of kinship with place and the desire to make connections with place are the offset to living amid nothingness (as social void). Taken up in the literature on ruination, deindustrialisation, urban decline, community collapse and the effects of 'natural' and 'human induced' disasters in place is the rippling effect of severing people's kinship with place. In such troubling moments as these, and indeed during ethnic conflict and violence, places may be forced into decline, through their transformation into battlefields, sites of wounding and hardship, even death. Places held close by an ethnic group may become the harbourer of poisons and toxic chemicals, the cause of constraint (limits to mobility, rights and freedoms) and may facilitate detainment (or imprisonment). When violence occurs in place, then what was meaningful, knowable and ordered by an inherent place logic can become disordered and harmful. How those who assert kinship with such transformed places might come to terms with this eventuality is a necessary point of reflection in this chapter, as it provides an opportunity to consider the fate of socially disordered places and some of the ways they might be brought back into a meaningful and nested ecology.

Making harm of beauty

Harmful places are the spaces of terror, humiliation and suffering that the wounded are forced to participate in. Harmful places are brought into existence or carved from the existing place world. Hence, their production may involve the transformation of home, community or nation into a site of harm. Good places may be transformed into bad ones. This undertaking threatens to break the kinship that is shared between people and place and may take generations to heal. In places where major disorder, mass death and destruction have occurred, withdrawal or evacuation may form part of a distancing in the relationship between people and place. Sometimes people will return to their place kin, but history attests also to the fact that many may not. Partly in the service of prescriptive forgetting, in order to survive, this distance may also reveal something of the trauma that comes from looking upon place in its disordered form.

Written of as the forging of harm from beauty – or in reverse, as the making of peace in a place of violence – this aspect of place disorder and reorder involves a deeply felt axiological crisis and attempts to remedy it. In the first instance, it involves the instating of sadness in place, planted so deeply that the joy of home, or the comfort and securities found through kinship with place become extremely difficult to locate. In the case of the latter, the axiological crisis is confronted and the intimacy that links people to place demands that place be recast in such a way as to affirm identity and bring strength, rather than detract from it. One part of what facilitates both experiences, or attempts to remedy social disorder in place, is what Connerton (2008) describes as prescriptive forgetting. This form of forgetting is most often an act of the state, yet here it is applied in broader terms to the actions of an ethnic or cultural group that comes to navigate disorder in place (Connerton 2008: 61). Prescriptive by nature, it is deliberate in its public acknowledgement of the dangers that come with remembering. Remembering past wrongs may lead to vendetta or, in the context of this discussion, a perpetual disordering of the place world and in turn human life. This caution acknowledges the fragility that conditions of social disorder can create.

Connerton (2008: 62) writes: "Sometimes at the point of transition from conflict to conflict resolution there may be no explicit requirement to forget, but the implicit requirement to do so is nonetheless unmistakable." This form of forgetting is timely in a discussion of place disorder, and the desire to smooth over the disarray that can be harboured in place in the aftermath or midst of ethnic violence, conflict and resulting cultural wounding. I write of this from the perspective of those people and places that are wounded, not those that perpetrate the violence. Accounts of intercommunal and ethnic violence show that home and parts of the place world once held dear can become sites of harm and violence. This is illustrative not only of place as the stage for conflict but also highlights that place can be operationalised for harm in the project of cultural wounding. To embed violence in an ethnic group's immediate world, and to transform places of security into ones of threat and uncertainty is a powerful way to undermine the esteem and psychology of a group. This form of chaos is designed to destabilise identities in the places where they find their strength and moorings. This often leaves in its wake a scarred landscape, made up of places that have been brutalised and rescripted through transformation.

In some instances, ethnic groups or communities that survive the violence that occurs in their place world may come to abandon home or retreat from territories, reflecting a type of prescriptive forgetting aimed at survival and necessitated healing. Where horrific events have occurred in familiar and everyday places, such as villages, homes, even on sporting fields, in marketplaces or at places of worship, the trauma may be held to linger in place, thus becoming capable of enacting a form of enduring and corrosive violence against ethnic constituents. Abandonment and refusal to return in the foreseeable future may be the only option for those who survive, as witnessed in places such as Aleppo in Syria, Luhansk in the Ukraine, and also across Palestinian villages in the West Bank and Gaza Strip. Ethnically cleansed villages in Chechnya, Kosovo and Bosnia-Herzegovina, where neighbourhoods and villages became sites of mass death during periods of war in the 1990s, also remain disordered in the minds of many survivors. Writing of ethnic cleansing, memory and homecomings in Bosnia and Herzegovina, Halilovich

(2011: 43), reflects that "[i]n the absence of a 'homeland' … memory – of not only *who I once was*, but also *who I now am* – often becomes the quintessential aspect of the individual and group identity of the displaced". Place, as the anchorage point for the vestiges of social identity, "gets compensated for – or 'kept alive' – by the memories and stories of the place lost" (Halilovich 2011: 43). In essence, place comes to hold strong through its very absence.

In a heart-wrenching account of witnessing violence in a safe place, Halilovich tells Edita's story. Edita is a Bosnian woman who survived the 1992 massacre of family members and neighbours in the village of Hegići. Halilovich (2011: 44–45) reports that many of those who survived the massacre, mainly women and children, were expelled from Hegići, and have never returned. Some 15 years after the massacre, now settled in Austria, Edita makes her 'homecoming' "to attend the burial of her father and grandfather, whose remains were found in a mass grave" (Halilovich 2011: 48). Describing the place to which Edita returns, Halilovich (2011: 49) paints an incredible picture of place, and the violence that is negotiated by seeking a return. As survivors look upon the ruins of their village, it is as though time has stopped, "and everything that had happened was still happening to them and their village" (Halilovich 2011: 49). Yet they "felt a strong sense of belonging to each other and to their dead relatives as well as a sense of moral superiority … After all, they did not kill anyone and they were burying *their* dead in *their* land" (Halilovich 2011: 49, emphasis in original).

Where physical return is not possible, or is counter to the emotional well-being of an ethnic group, it is often stories that become vessels for holding and keeping places (Halilovich 2011: 44). "We are able to root our sense of place and our sense of self through story" (Halilovich 2011: 44). The spirit of place can be upheld through memory and story, or it may be upheld by a degree of prescriptive forgetting, where the worst elements of trauma in place are not allowed to prevail in its ongoing narrative. This process is complicated and requires consciousness of past events and future needs, which temper the role of remembering and forgetting in the enduring relationship people have with place. Halilovich (2011: 44–45) describes the interplay between places and stories – or memories of places – especially with those places that are "vandalized, butchered, divided up and scarred". Place, at once "evoked, narrated and remembered, can [also] become a constant reminder of humiliation and suffering" (Halilovich 2011: 45). Spaces such as these, confused and contested, may fill or be filled with "trauma that triggers psychic and visceral reactions on the part of expellees, but it also remains a place of desire, with those who once lived there propelled by a sense of obligation to keep telling its stories" (Halilovich 2011: 45).

Prescriptive forgetting in these instances is designed to overshadow the effects of enduring suffering caused by place harm and can operate to heal place by reinstating and witnessing its former glory. It preferences survival and delimits the pathways to enduring suffering. The deliberate forgetting of harms and violence endured in place hopes to return order to place, acknowledging it as a benevolent presence, and a source of strength. The focus is on life over death, with place being key to the narration of a culture that outlived the experience of hardship or survived wounding events. Where new places have been created and introduced into the place world of an ethnic group, these too must be somehow accommodated into the ongoing narrative of people in place.

Introduced places, which come about as part of a wider project of cultural wounding, may include those to which ethnic groups are dispatched through forced removal, prison camps or other sites of detainment, other countries where the persecuted are forced to flee as legal or illegal immigrants, slave markets, work camps or reservations. Places such as these may enter into a meaningful place world for an ethnic group, depending on how key a role they play in group and individual survival. Places of harm can be factored back into the place world, their damaging qualities made into a strategic resource that feeds people in their journey to healing. This is evidenced by memorials that provide reflection in places where atrocities have occurred, and by the reclaiming of sites of suffering as spaces for identity affirmation and performance (Margry and Sanchez-Carretero 2011; Staub 2000: 379).

In times of war, ethnic violence and conflict, places are produced and scripted into the wounding enterprise. This is seen in the case of the transatlantic slave trade, a practice that had at its centre a motive for violence and a denial of rights to certain ethnic groups. This can be read as an historical practice embedded in ethnic violence, where, "ethnic identities served to define a category of 'others' who were legitimately enslavable" (Law 1997: 205). Coastal parts of West Africa were turned into slave ports. Key regions, which became known as the Slave Coast and the Gold Coast, reinscribed coastal parts of West Africa and the Bight of Benin as places of suffering for those who were captured, enslaved and sold. The Slave Coast was a major source of African slaves during the Atlantic slave trade, from the early sixteenth century to the late nineteenth century. Ports along the coastline from where the enslaved were dispatched to foreign shores, include Ouidah (Benin), Lagos (Nigeria), Aného (Togo), Grand Popo (Benin), Agoué (Benin), Jakin (Benin), Porto-Novo (Benin) and Badagry (Nigeria) (Law 1991; Lovejoy 2012: 54–56; Rodriguez 1997). For those captured, sold and transported across the Atlantic, and certainly for those who arrived in Brazil, the place world of enslavement, as stretched across the Diaspora, has become a cornerstone of black rights movements and African Brazilian identity affirmation (see Araujo 2014, 2015). Today it is recognised as having substantial heritage value for many African Brazilians and black rights groups, for it is the world that contains the experiences of their ancestors, as those who were taken from their homes and thrust into a world of traumatic encounters and associated place-making. Today this heritage is enlivened through the rhetoric of journeying, Diaspora and the vast place distinction of Mama Africa.

The initial capture and transportation of people from their tribal homes was a violent and disordering event that turned the known place world into the unknown for their descendants. While some people enslaved in Brazil returned home in the years after emancipation, for many others, Africa as home became a diffused notion; their ancestral home existing only in relation to life in Brazil and the journey that took them there. Mama Africa was made to harbour the bodies of the enslaved as they awaited transportation from key sites along the Slave Coast. They then began their journey through an unkind place world in which survival was not assured. This world was demarcated by slave ports, barracoons (slave sheds), 'gates of no return', vessels at sea in which slaves were relinquished below deck in harrowing conditions, markets where bodies were traded and then fazendas where those enslaved were compelled to work in harsh conditions. From this came a new place world, which today invokes African nations, the once familiar and now imagined villages of ancestral homes and the places that slaves passed through in their time of captivity. It also houses places of resistance, hiding and escape. This is an extremely complex universe of places, one that calls upon Mama Africa to distinguish its boundaries. It is a vast place world united through experiences of safety and harm and the ongoing transformation of people and place-meaning.

As horrific as the histories attached to these places – and therefore their place identities – might be, many are invoked in survival narratives and have become key points across the African Brazilian political and social landscape. They have even entered into a network of places packaged together for 'cultural roots' tourism and have become sites of healing for persons from all across the African Diaspora. A relatively recent addition to this network of places in Brazil is the Valongo Wharf, in Rio de Janeiro. Uncovered by way of excavation work in 2011 (in preparation for the 2016 Olympics), this place is one part of a complex of sites, including a group of warehouses used for quarantining, displaying and selling enslaved Africans brought to Brazil, a lazaretto (a place for the treatment of contagious diseases) and a cemetery (the Pretos Novos Cemetery). Since its excavation, African Brazilian organisations have begun organising heritage tours, public religious ceremonies and capoeira performances in the Valongo area. This, at a place where more than 1 million Africans came ashore as enslaved persons (Araujo 2014: 101; Cicalo 2015).

Elsewhere in Brazil, the Mercado Modelo, located in the lower city of Salvador Bahia has become heavily associated with slave disembarkation and contemporary expressions of blackness and African descent. Although not an actual site of disembarkation, yet located close by to the site where slaves arrived, the Mercado Modelo has become a vital hub in the place world of Brazilian black identity in the northeast.[5] Today it is host to public performances of African Brazilian arts, music and martial arts (capoeira). African Brazilians perform their identity here and affirm links back to the ancestral experience of slavery, disembarkation and survival in Brazil. So too the 'objects' associated with this identity are available through the consumable tourist items sold in the marketplace. Araujo (2014: 103) reflects that the legend of this place as one of disembarkment and slave suffering is a "case of memory replacement", in a country where "the local Afro-Brazilian population deals with a lack of visible and official markers indicating the existence of sites remembering the Atlantic slave trade in Salvador".

Another potent place that is today scripted into the prevailing narrative of the African Diaspora in Brazil is that of the Quilombo dos Palmares. This was a large settlement founded by escaped slaves from around the late 1500s. It is believed to have been spread across the region of northern Alagoas and southern Pernambuco (Anderson 1996: 551) and is considered a holy place for black identity and politics in Brazil (Anderson 1996). Once home to more than 30,000 escaped slaves, the Quilombo dos Palmares faced multiple attacks from colonial and pioneering forces, yet retained a strong resistance in large part due to the leadership of freedom fighter Zumbi dos Palmares. Today, both the Quilombo and Zumbi dos Palmares are enshrined in narratives of the African Diaspora, through which the history of this place and the wider region channels, as inspiration for black activists who campaign for liberation, human rights, dignity, respect and equality. The contemporary reclamation and transformation of sites of deportation and disembarkation, as well as slave markets and quilombos, is a testament to the making of beauty and strength in places where harm and cultural wounding found and took their hold.

Heritage sites such as these are spread across the African Diaspora in Senegal, Ghana and the Republic of Benin, where persons of varied ethnic backgrounds passed through 'gates of no return'. Today their descendants and other members of the African Diaspora visit such places to remember the journey their ancestors were forced to make. "For many decades the sites of arrival remained concealed in the public spaces of former Atlantic slave ports because most old port areas had become either abandoned, impoverished, or replaced with new construction" (Araujo 2014: 76). "In former slave ports like Rio de Janeiro and New York, the unearthing of slave wharfs and slave cemeteries over the last 20 years is finally leading public authorities to formally establish permanent makers to commemorate Atlantic slavery" (Araujo 2014: 213). The African Burial Ground in New York City, with its pronounced burial mounds and monument to those who died as enslaved persons, is an active intervention into any erasure of this violent past and its long-term culturally wounding effects in the United States. As these places are revealed, they enter into the corpus of heritage and reinstate themselves as part of the place world that is the African Diaspora. In Brazil, places once laden with misery and the suffering of slaves become strategic resources in the affirmation and declaration of an African Brazilian identity in place and time. It is as if a bargain has been struck by the descendants of those enslaved and place, that place-meaning needs to be given an ordered sense within the narrative of survival so as to nurture the identities of those once harmed and those who survived. This does not constitute the erasure of memory associated with trauma in place, but instead stands to liberate place from its violent past. A type of prescriptive forgetting happens in this instance, where the horrors alone cannot be all that defines a place and its kinship with people. Kinship is then traced through the links between people and place, with place regarded as a necessary facilitator of survival.

Chapter overview

Social disordering as a form of place harm remains a pervasive threat in the world today. It is not limited to single sites, nor constrained by the event of direct conflict. Ethnic violence and, in turn, place harm send ripples out further, to incorporate a vast number of places, both new and old. This is the case with the place world that constitutes Mama Africa and the African Diaspora. It is also playing out nightly on our televisions screens as reports of drowned immigrants off the coasts of places such as Lampedusa, in the Mediterranean, challenge us to consider whose lives matter and which places might harbour these lives in safety. This small island is in the midst of social disorder, affected by a vast number of places where ethnic conflict and violence occur. As the closest European territory to Libya, it has become a key transit point for those seeking to flee war-torn territories and ethnic violence in Africa, the Middle East and Asia. People fleeing ethnic violence risk their lives aboard unseaworthy vessels in the hope of escaping violence and its impending trauma. As the bodies of these desperate refugees wash ashore, the very identity of Lampedusa and in turn the whole of Europe is changing. Place is not impervious to the effects of this violence and the cultural wounding it induces. As one place finds itself emptied of ethnic constituents and another is confronted by the loss of life and the traumas this brings, a deeply wounded space is born.

This chapter has sought to convey aspects of social disorder in place, prompting reflection on the pervasiveness of this experience. Social disorder has a profound effect on human life and, as emphasised here, it also explicitly effects place. For many ethnic groups, the social order of everyday life is tied closely to the order of place. The kinship between people and place holds in balance, or knowable accord, an entire form of existence. To attack one part of this, namely the place world, is a proven and effective method for undermining ethnic identity and human security. The effects spread further and aggressively change place, which is denied its own sentiency and stripped of existing order. Social disorder is distinguished here as that a form of place violence instated by the erasure of toponymic distinctiveness and renaming, the denial of local place empiricism and order, the erasure of narratives and testimonies of such place order and by the making of harmful places.

Social disorder is a form of place harm that may or may not leave tangible traces, thus it is distinct from the destruction and designification of place as discussed in Chapter 2. It does, however, form part of an arsenal of place harming strategies that achieves chaos in the hearts and minds of those who identify closely with place. So too it attacks the identity of place itself. It is akin to the tearing up of place, as if it were a blank piece of paper, torn to pieces and then scattered into the wind. Where those pieces fall is often of little consequence to the perpetrators of violence as agents of harm, for they plan to rewrite place in their own fashion – through new names, new stories and a whole new axiology. Where those torn pieces fall, however, will be of profound importance to the human kin that constitute place and are constituted by place. When aspects of your DNA are under threat and face erasure, then the emphasis is on locating the pieces and striving to put things back together again. There are, however, many obstacles to this being done and many different approaches may be taken to reinstate kinship and recovering place-meaning.

This chapter is intended to trigger reflection of the effects of ideological and emotional violence against place. That this disorder can tip over into chaos and death is highly likely, and this becomes the focus of discussion in the following chapter. Chapter 5 introduces the theme of elemental erasure, as a third strategy of place harm. Examining elemental erasure involves looking closely at the parts that make up place. Recognising place for both its tangible and intangible qualities, these elements are held to be the ecology of place, holistically viewed as the rich tapestry of non-human life and otherly agents that coexists in place. Co-presence is an often-returned-to theme in this book, and works to articulate the existence of multifarious agents, beyond but inclusive of the human, in place. When any element of this is the target of violence or recipient of harm, then this may seep into the structure of place, potentially expanding disorder into a chaotic state, akin to toxicity, decay and death.

Notes

1 Elemental destruction is discussed further in Chapter 5, with particular reference to ecological decline and the effects of disorder and destruction on place's constitutive parts. Elemental destruction can refer to the loss of floral and faunal life, the introduction of feral species, pollution and contamination, as shifts that come as place identity and order are denied by powerful others or place invaders. While it is commonly held that the key elements of place are those of physical distinction, in a biological and geological sense, it may also be the intangible aspects of place that make up its constitutive parts. Intangible elements include ancestral spirits that move through place, the presence of a particular language among human kin, to which place, as a sentient being, responds and other communicative pathways (such as ceremonial or religious order) that place requires to commune with kin.

2 Transfer names are those that repeat British nomenclature (Atchison 1990: 154).

3 Renaming creates new connections and often obliterates old ones. Alderman (2008: 195) even considers the role whites played in naming African slaves. Many of those enslaved kept the names of slave masters long after emancipation, while others challenged the violence of their origins, such as Malcolm X, refusing the namesake of 'Little' as granted to his ancestors and using X to reference his stolen and unknown tribal name that marked his origins (Alderman 2008: 195, citing Baugh 1991).

4 The meeting occurred in April 2003, the same month in which the Turkish Cypriot administration significantly eased travel restrictions across the Green Line, allowing Greek Cypriots to cross at the Ledra Palace Crossing.
5 Araujo (2014: 103–104) recounts the journey by which the Mercado Modelo incorrectly became regarded as one of the largest historical ports for the arrival of enslaved persons from Africa. It is such that, "the present-day building of the Mercado Modelo was not the actual central slave market, but rather that one was located in another site close to the current location". The original building was destroyed by fire in 1969 (Araujo 2014: 13).

References

Alderman, D. 2008. Place Naming and the Interpretation of Cultural Landscapes, in *The Ashgate Research Companion to Heritage and Identity*. Edited by B. Graham and P. Howard, pp. 195–212. Aldershot, Hampshire: Ashgate Publishing.

Anderson, R.N. 1996. The Quilombo of Palmares: A New Overview of a Maroon State in Seventeenth-Century Brazil. *Journal of Latin American Studies*, Vol. 28(3): 545–566.

Araujo, A.L. 2014. *Shadows of the Slave Past: Memory, Heritage, and Slavery*. New York: Routledge.

Araujo, A.L. (ed.) 2015. *African Heritage and Memories of Slavery in Brazil and the South Atlantic World*. Amherst, New York: Cambria Press.

Arefi, M. 2007. Non-Place and Placelessness as Narratives of Loss: Rethinking the Notion of Place. *Journal of Urban Design* Vol. 4(2): 179–193.

Atchison, J. 1990. Naming Outback Australia. *Proceedings of the XVIth International Congress of Onomastic Sciences*, Quebec: Universite Laval, 16–22 August 1987.

Auge, M. 1995. *Non-Place: Introduction to an Anthropology of Super Modernity*. New York: Verso.

Bakshi, A. 2012. A Shell of Memory: The Cyprus Conflict and Nicosia's Walled City. *Memory Studies* Vol. 5(4): 479–496.

Basso, K. 1988. 'Speaking with Names': Language and Landscape Among the Western Apache. *Cultural Anthropology* Vol. 3(2): 99–130.

Basso, K. 1996. *Wisdom Sits in Places: Landscape and Language Among the Western Apache*. Arizona: University of New Mexico Press.

Baugh, J. 1991. The Politicization of Changing Terms of Self-Reference Among American Slave Descendants. *American Speech* Vol. 66(20): 133–146.

Berg, L. and Kearns, R. 1996. Naming as Norming: 'Race', Gender and the Identity Politics of Naming Places in Aotearoa/New Zealand. *Society and Space* Vol. 14(10): 92–122.

Borowiec, A. 2000. *Cyprus: A Troubled Island*. Westport, Connecticut: Praeger Publishers.

Brattland, C. and Nilsen, S. 2011. Reclaiming Indigenous Seascapes: Sami Place Names in Norwegian Sea Charts. *Polar Geography* Vol. 34(4): 275–297.

Bryant, R. 2012. Partitions of Memory: Wounds and Witnessing in Cyprus. *Comparative Studies in Society and History* Vol. 54(2): 332–360.

Cajete, G. 1999. Look to the Mountain: Reflections on Indigenous Ecology, in *A People's Ecology: Explorations in Sustainable Living*. Edited by G. Cajete, pp. 3–20. Sante Fe: Clear Light Publishers.

Carter, P. 1988. *The Road to Botany Bay: An Exploration of Landscape and History*. New York: Alfred A. Knopf.

Casey, E. 1996. How to Get from Space to Place in a Fairly Short Stretch of Time: Phenomenological prolegomena, in *Sense of Place*. Edited by S. Feld and K. Basso, pp. 13–52. Santa Fe: School of American Research Press.

Cicalo, A. 2015. From Public Amnesia to Public Memory: Re-Discovering Slavery Heritage in Rio de Janeiro, in *African Heritage and Memories of Slavery in Brazil and the South Atlantic World*. Edited by A.L. Araujo, pp. 180–211. New York: Cambria Press.

Connerton, P. 2008. Seven Types of Forgetting. *Memory Studies* Vol. 1(1): 59–71.

Council of Europe. 2001. *Cyprus v. Turkey (Application Number 25781/94)*, European Court of Human Rights. Available at www.refworld.org/cgi-bin/texis/vtx/rwmain?docid=43de0e7a4. Accessed 31 August 2015.

Creative Spirits. nd. *Racism in Aboriginal Australia*. Available at www.creativespirits.info/aboriginalculture/people/racism-in-aboriginal-australia#toc9.

Davidson, J., Bondi, L. and Smith, M. (eds). 2007. *Emotional Geographies*. Aldershot, Hampshire: Ashgate Publishing.

Deur, D. 1996. Chinook Jargon Placenames as Points of Mutual Reference: Discourse, Intersubjectivity, and Environment within an Intercultural Toponymic Complex. *Names* Vol. 44: 291–321.

Guyot, S. and Seethal, C. 2007. Identity of Place, Place Identities, and Change of Place Names in Post-Apartheid South Africa. *South African Geographical Journal* Vol. 89(1): 55–63.

Hagan, S. 2006. *Australia's Blackest Sporting Moments: The Top 100*. Darling Heights, Queensland: Ngalga Warralu Publishing.

Hagan, S. 2007. *Nigger Lovers*. Documentary. Produced by Stephen Hagan and Daryl Sparkes. Australian Film Commission, Film Finance Corporation, Pacific Film & Television Commission.

Halilovich, H. 2011. Beyond the Sadness: Memories and Homecomings Among Survivors of 'Ethnic Cleansing' in a Bosnian Village. *Memory Studies* Vol. 4(1): 42–52.

Hercus, L. and Simpson, J. 2002. Indigenous Placenames: An Introduction, in *The Land is a Map: Placenames of Indigenous Origin in Australia*. Edited by L. Hercus, F. Hodges and J. Simpson, pp. 1–23. Canberra: Australian National University E Press.

Herman, R.D.K. 1999. The Aloha State: Place Names and the Anti-Conquest of Hawaii. *Annals of the Association of American Geographers* Vol. 89(1): 76–102.

Ingold, T. 2000. *Perception of the Environment: Essays on Livelihood, Dwelling and Skill*. Abingdon, Oxon: Routledge.

Kearney, A. 2014. *Cultural Wounding, Healing and Emerging Ethnicities: What Happens when the Wounded Survive?* New York: Palgrave Macmillan.

Kearney, A. and Bradley, J. 2009a. Too Strong to Ever Not Be There: Place Names and Emotional Geographies. *Journal of Social and Cultural Geography* Vol. 10(1): 77–94.

Kearney, A. and Bradley, J. 2009b. Manankurra: What's in a Name. Place Names and Emotional Geographies, in *Aboriginal Placenames Old and New: Discovering, Interpreting and Restoring Indigenous Nomenclature for the Australian Landscape*. Edited by H. Koch and L. Hercus, pp. 463–480. Canberra: Australian National University E Press and Aboriginal History Inc.

Kovras, I. 2012. Explaining Prolonged Silence in Transitional Justice: The Disappeared in Cyprus and Spain. *Comparative Political Studies* Vol. 46(6): 730–756.

Law, R. 1991. *The Slave Coast of West Africa 1550–1750: The Impact of the Atlantic Slave Trade on an African Society*. Gloustershire: Clarendon Press.

Law, R. 1997. Ethnicity and the Slave Trade: 'Lucumi' and 'Nago' as Ethnonyms in West Africa. *History in Africa* Vol. 24: 205–219.

Loizos, P. 1975. *Greek Gift: Politics in a Cypriot Village*. Oxford: Oxford University Press.

Lovejoy, P. 2012. *Transformations in Slavery: A History of Slavery in Africa*. Cambridge: Cambridge University Press.

Magudu, S., Muguti, T. and Mutami, N. 2010. Political Dialoguing Through the Naming Process: The Case of Colonial Zimbabwe since 1890. *Journal of Pan African Studies* Vol. 3(10): 16–30.

Magudu, S., Muguti, T. and Mutami, N. 2014. Deconstructing the Colonial Legacy through the Naming Process in Independent Zimbabwe. *Journal of Studies in Social Science* Vol. 6(1): 71–85.

Margry, P. and Sanchez-Carretero, C. 2011. *Grassroots Memorials: The Politics of Memorializing Traumatic Death*. New York: Berghahn Books.

Morphy, H. 1995. Landscape and the Reproduction of the Ancestral Past, in *The Anthropology of Landscape: Perspectives on Place and Space*. Edited by E. Hirsch and M. O'Hanlon, pp. 184–209. Oxford: Clarendon Press.

Munn, N.D. 1996. Excluded Spaces: The Figure in the Australian Aboriginal Landscape. *Critical Inquiry* 22(3), pp. 446–465.

Nash, C. 1999. Irish Placenames: Post-Colonial Locations. *Transactions of the Institute of British Geographers* Vol. 24(4): 457–480.

Navaro-Yashin, Y. 2002. *Faces of the State: Secularism and Public Life in Turkey*. Princeton: Princeton University Press.

Navaro-Yashin, Y. 2005. Confinement and the Imagination: Sovereignty and Subjectivity in a Quasi-State, *Sovereign Bodies: Citizens, Migrants and States in the Post Colonial World*. Edited by T.B. Hansen and F. Steputtat, pp. 103–119. Princeton: Princeton University Press.

Navaro-Yashin, Y. 2006. De-Ethnicizing the Ethnography of Cyprus: Political and Social Conflict between Turkish-Cypriots and Settlers from Turkey, in *Divided Cyprus: Modernity and an Island in Conflict*. Edited by Y. Papadakis, N. Peristianis and G. Welz, pp. 84–99. Bloomington: Indiana University Press.

Navaro-Yashin, Y. 2009. Affective Spaces, Melancholic Objects: Ruination and the Production of Anthropological Knowledge. Malinowski Memorial Lecture. *Journal of the Royal Anthropological Institute* Vol. 15(1): 1–18.

Navaro-Yashin, Y. 2010. The Materiality of Sovereignty: Geographical Expertise and Changing Place Names in Northern Cyprus, in *Spatial Conceptions of the Nation: Modernizing Geographies in Greece and Turkey*. Edited by P.N. Diamandouros, T. Dragonas and C. Keyder, pp. 127–143. London: I.B.Tauris.

Navaro-Yashin, Y. 2012. *The Make-Believe Space: Affective Geography in a Post-War Polity*. Durham: Duke University Press.

Papadakis, Y. 2000. The Social Mapping of the Unknown: Managing Uncertainty in a Mixed Borderline Cypriot Village. *Anthropological Journal on European Cultures* Vol. 9(2): 93–112.

Papadakis, Y., Peristianis, N. and Welz, G. 2006. Introduction, in *Divided Cyprus: Modernity, History, and an Island in Conflict*. Edited by Y. Papadakis, N.

Peristianis and G. Welz, pp. 1–30. Bloomington, Indiana: University of Indiana Press.

Roberts, J.T. 1993. Power and Placenames: A Case Study from the Contemporary Amazon Frontier. *Names* Vol. 41(3): 159–181.

Rodriguez, J. 1997. *The Historical Encyclopedia of World Slavery*. Denver, Colorado: ABC CLIO, University of Michigan.

Rose, D.B. 2004. *Reports from a Wild Country: Ethics for Decolonisation*. Sydney, New South Wales: UNSW Press.

Salmón, E. 2000 Kincentric Ecology: Indigenous Perceptions of the Human-Nature Relationship. *Ecological Applications* Vol. 10(5): 1327–1332.

Sant Cassia, P. 2005. *Bodies of Evidence: Burial, Memory and the Recovery of Missing Persons in Cyprus*. New York: Berghahn Books.

Staub, E. 2000. Genocide and Mass Killing: Origins, Prevention, Healing and Reconciliation. *Political Psychology* Vol. 21(2): 367–382.

Tamisari, F. 2002. Names and Naming: Speaking Forms into Place, in *The Land is a Map: Placenames of Indigenous Origin in Australia*. Edited by L. Hercus, F. Hodges and J. Simpson, pp. 87–102. Canberra: Australian National University E Press.

Webber, M. 1964. The Urban Place and the Non-Place Urban Realm, in *Explorations into Urban Structure*. Edited by M. Webber, J. Dyckman, D. Foley, A. Guttenberg, W. Wheaton and C.B. Wurster, pp. 79–153. Philadelphia: University of Pennsylvania Press.

Yeoh, B.S.A. 1992. Street Names in Colonial Singapore. *Geographical Review* Vol. 82(3): 313–322.

Yeoh, B.S.A. 1996. Street-Naming and Nation-Building: Toponymic Inscriptions of Nationhood in Singapore. *Area* Vol. 28(3): 298–307.

5 Elemental erasure and ecological decline

Further advancing the diagnosis of place harm induced by violence and cultural wounding, this chapter moves to consider elemental erasure and ecological decline. These involve the delivery of harm to place's constitutive parts, as a deliberate strategy to erode the foundations of and relational setting for human life. This comes as a consequence of ambivalence to and disregard for place value. The logic of death from below, and all around, is central to elemental erasure and ecological decline as forms of place harm and wounding. Elemental erasure refers to the loss of life, inclusive of non-human elements: the flora, fauna and all those dynamically interacting organisms and the communities they form that constitute living presences in place. The killing of non-human life in place and the decline of ecosystems in the midst or aftermath of conflict may trigger destructions and social disorder, if not place death. The killing of Indigenous species, the introduction of feral ones and the instating of pollution and contamination are often the result of ambivalence towards place value and are reflective of deep axiological crises, which come as place identity and order are denied by powerful others and place invaders. These are but some of the weapons of actual and ideological warfare in conflict zones, where the impetus is to drive out ethnic groups from place, by rendering place uninhabitable, or to disrupt the ecology of place and instate sickness that may eventually leach into the lives of place's human kin.

Ecosystem death and destruction are often key strategies in campaigns of cultural wounding and attacks on ethnic groups, and can bring about substantial psychological unrest and fear for those who depend upon place, and formulate relations to place through kinship. During episodes of ethnic conflict, it is as Heider (2005: 12) writes, in an account of 'violence and ecology': "Violent action means usually producing spectacular changes from order to disorder, which can be brought about by undirected or only roughly directed launching actions." Deep unrest comes for those ethnic groups who are bound closely to place and homelands, and for them, decline may be read as a form of communicative event, in which kin are charged with somehow relinquishing their responsibility. In an attempt to understand the decline of place, people seek epistemological, ontological and axiological structures to make sense of what they are witnessing. If the relationship with place is one defined by intimacy and kinship, then place decline may reflect a breakdown in relations or a breach of the principles that govern this relationship. In turn, people may trace decline and loss to negligence on their behalf, and more tragically to the neglect of their place kin.

The expression 'might be something', often utilised by Indigenous Australians, speaks to the translation of meaning around indeterminate or unpredictable events as seen across homelands. It is often uttered throughout discussions of place health and ill health, revealing an uncertainty as to what is going on in place, while hinting at the possibility of human responsibility (Kearney and Bradley 2015; Povinelli 1993). The uncertainty of change in place and its indeterminate cause is grappled with by way of a search for meaning, conveyed through the exclamation of 'might be something'. This refers to the possibility that when events occur and change appears the most varied, fates may collide and interweave with one another. This language of indeterminacy (see Povinelli 1993) speaks to the relational quality of human presences in country and also to the pervasive vitality and power that resides in all elements of life; from kin, country, Law, ancestors and the natural elements and phenomena that make up country (Kearney and Bradley 2015: 175). At any point these elements can interact with one another in ways that might convey imbalance, misuse of power or worrisome actions and consequences.

Writing in 1993, Povinelli introduced readers to 'sign reading' among Belyuen women and Indigenous Australian cultures more broadly. She (1993: 687) unpacked the work of Stanner (1965: 217, in Povinelli 1993: 687) that concerned itself with "Australian Aborigines' 'obsessive pre-occupation' with the 'signs, symbols, means, portents, tokens, and evidences of vitality' in the environment", remarking that Indigenous people "do not, however, see every overturned stone as manifesting Dreaming intentionality … [and indeed] there were many things in the environment that were just things, themselves only and no more, without import, standing for nothing" (Stanner 1965: 217, in Povinelli 1993: 687). The ability to discern one from the other is a sign of intelligence and insight, as it requires knowledge, mindfulness and full activation of kinship with place. The need to read the signs, such as anomalies of species shape, behaviour and location, shifts in elements and natural phenomena; is associated with risk management, reciprocity, social action, politics, ecological probability and happenstance (Povinelli 1993: 684). Being observant of behavioural nuances and the ability to detect subtle meanings and thus generate "a meaning-claim based on a longstanding experience of the social and cultural landscape is a particularly persuasive strategy that holds the hegemonic higher ground" in Indigenous communities (Povinelli 1993: 693). The observation of 'might be something' and the need to interrogate unpredictable encounters is expressed further,

> The country is like the body in its capacity to express numerous social, mythic and historical dramas and in its necessity to be watched and interpreted ... Just as the country is constantly emitting signs, so people are constantly displaying their prowess or clumsiness in interpreting them.
>
> (Povinelli 1993: 692)

Kinship demands that the first point of reflection is the effect of human behaviour and action. From here, steps are taken to trace back to the source of harm. This process induces stress, particularly among Indigenous elders who hold strong to the Law of kinship and being responsible for country, and even shame among younger generations who may feel ill-equipped to enact the necessary relationships to remedy harm and suffering in place. The great burden of these emotive states and the emotional geographies they induce has been an enduring part of the settler colonial experience for Indigenous families in Australia. Their homelands have been the colonial object of desire and this desire has fed an insatiable need for the settler population to possess place or at least heavily designify and resignify it. This has been unrelenting, and in recent years is most powerfully expressed through the activities of the natural resource boom and mining industry, which has been unparalleled in its capacity to possess and harm, if not destroy place.

Shkilnyk (1985) documents similar practices of translating meaning in moments of place chaos caused by toxicity. She writes of the events that lead to the relocation of the Ojibwa people, the Indigenous owners of the islands and peninsulas on the English-Wabigoon river, northwestern Ontario, Canada, and the ongoing effects this move had on members of the community (Shkilnyk 1985: 2). The Ojibwa were moved from their homelands, only to then experience a mercury poisoning that occurred in waterways on the new reserve. This triggered a decline in community life, health and well-being, and Shkilnyk's (1985) ethnography recounts Ojibwa testimonies of 'coming to terms' with what has occurred. Many traced the events of mercury poisoning to ill-ease within the new reserve. As a sentient co-presence, this place was held to be "bad land", "off limits to human habitation" and full of "bad spirits" (Shkilnyk 1985: 70–71). The resounding sentiment was that this was not a place for Ojibwa, and that place was communicating loudly through ominous warnings such as persons and spirits surfacing from the lake, poor quality of light in place and the presence of "troubled spirits" on the new reserve. Such determinations reflect Ojibwa Law and were "perceived by the inner senses, the 'eyes of the soul'"; communicated by place to those with the kinship, and eyes (cultural vision) to see (Shkilnyk 1985: 70–71). Writing of the harrowing effects of toxicity and contamination, in the aftermath of relocation and community decline, Shkilnyk reflects:

> Having just been wretched from their moorings on the old reserve, the people were ill prepared to cope with yet another misfortune. They had but a precarious hold on the conditions of their existence on the new reserve. They could no longer draw strength either from their relationship to the land or from the well of their faith, which had once given meaning and coherence to their lives. In the context of the traditional religious beliefs, the contamination of the river could only be interpreted as punishment by the Great Spirit for some serious violation of the laws governing man's relationship to nature. People had great difficulty comprehending this 'unseen poison' of mercury, whose presence in the water and in the fish they could not see or taste or smell.
>
> (Shkilnyk 1985: 179)

Attempts to understand elemental erasure and ecological decline in place can convey desperation and a searching through epistemological frameworks to identify structures of knowing and explanation. The determination of responsibility is a heavily negotiated undertaking. In the long run, agents of harm may be identified as persons and processes beyond the ethnic group, but uncertainty can cause members of the group to look back critically at themselves, a condition that can lead to shame and associated hardship in the lives of those deeply attached to place. Simultaneously erasure and decline eat away at the well-being of people and place, to the point where suffering is a co-terminus event.

Elemental erasure

Geography, concerned with elements of the place world, spatially distinguishes relationships between people, places, and environments by mapping information about them. It strives to show how identities and lives of individuals and peoples are rooted in particular places and examines how these places come to be, as shaped by physical processes and as interactions between plants, animals and overall ecosystems. Placing the human centrally within this interacting universe is what defines the discipline's humanist turn. Geography is in whole a discussion of interacting elements. This elemental preoccupation can also be explored, relative to place harm and the greater effects of cultural wounding. Operating along lines of relational influence, the notion of elemental erasure is inspired by geography in this broad sense, as well as the more discrete principles of nesting and nested ecologies.

Evidenced by ecocide, as the deliberate or consequential large-scale destruction of the natural environment, along with species decline and violent land-use practices that bring with them toxicity and pollutants, there is no denying the propensity for elemental erasure to precipitate place death by a million cuts (Harvey 2012; Higgins 2010).[1] By extending this discussion of violence and harm to include the experiences of the non-human, kinship between people and place is expanded to include all elements of place. This includes kinship with non-human species; implying a whole sequence of relationships expressed through familial connections, ancestral links and economic relations. It is neither radical, nor fantastical that, in accordance with an Indigenous epistemology, a white-bellied sea eagle can be one's maternal grandmother, the conch shell a paternal ancestor, the brolga one's father's father and the groper one's mother. This is the pervasive nature of kinship. These are substantive relationships that form the kinship structure of many Indigenous Australian groups (see Bradley 2008; Rose 1996, 2000, 2008c, 2013; Tamisari 2002), invoking responsibility, affinity and obligation.

Much of what is contained in the literature on ecology, and in particular that concerning Indigenous ecological perspectives, is a pervasive sense of relatedness, and connectivity. The common definition of ecology is the (scientific) study of interactions among organisms and their environment. Left at that, the very concept of ecology can be applied without limits. It has thus leant itself well to incorporation within the discourses of both geography and anthropology and expands its intellectual fit, today most energetically through environmental and ecological humanities, political economic ecology and even quantum anthropologies (Abel and Stepp 2003). In each case, the emphasis is on interactions and relationships, many of which are bound by principles of mutuality and response. In the context of this book, such principles of interaction and mutuality are adopted, relative to meanings these might carry in relation to place and cultural geography. This does not end with the recognition of place as part of the broader ecology, in a physical and tangible sense, but that place itself contains ecological order as the convening and communing presences that exist in place. These may be people, non-human animals, waterways, grasslands, soil, geology, ancestral beings, spiritual agents and otherly agents that coexist in place. So while place can be read as part of a wider ecology, so too contained within place are its own inherent sets of relations; its local empiricism. These are often key to the cultural and emotional geographies of place and are what become vulnerable to attack during episodes of ethnic conflict and cultural wounding.

By defining ecology as sets of relations, and then instating a cultural lens to the reading of this as kinship, I am encouraged to do as Rose (2014: 431–432) has done elsewhere, namely, 'sidestep nature'. It is now widely recognised that "not all humans inhabit the west's nature-culture/mind-matter binaries" (Rose 2014: 431–432). Thus best practice in any discussion of the relational pedagogy of place and the effect of its being harmed is to "situate humans within webs of life, thus radically undermining the nature/culture dualism that has held sway in the west for so long" (Rose 2008a: 158). In line with this, and out of a reluctance to replicate the West's notion of ecology as environmentally grounded, its status as the condition for hyper-relationality in place is brought to the fore, where separation between nature and culture agents is no longer sought, as this only functions to delimit the comprehension of place elements communicating and responding to one another. Drawn to a similar praxis, Muir, Rose and Sullivan (2010: 259) introduce a discussion of philosophical ecology as evidenced by Indigenous knowledge and the Darling River in southeastern Australia, by instating this form of hyper-relationality. They contend that "[a] river is like a mirror: it reflects the care given by people whose lives depend upon it". The decline of the riverine ecology "reveals more than troubled ecological relationships", and is believed to communicate the depth of ruptured social relations (Muir et al. 2010: 259).

The reference to 'mirroring' here is what captures my attention and returns this discussion to a reflection on substantive relations between people and place, as outlined in Chapter 1. By emphasising the importance of social relationships for good ecological relationships, the authors remind the reader of the epistemological and ontological structures that refute the possibility of separation between nature and culture, people and place (Muir et al. 2010: 259). Engaging this view further, they cite Sullivan, Indigenous co-author and contributor, for whom the Darling River is an ancestral home: "The environment is a reflection of who we are as human beings, and the environment is in a crappy way. And you know why it's in a crappy way? Because we're in a crappy way. That's the bottom line. The environment is terrible. The river is terrible" (Phillip Sullivan, 24 July 2008, Bourke, NSW, cited in Muir et al. 2010: 262).

Circles within circles is how place is imagined as made up of elements, as multifarious and interacting agents. Hence the appeal of nesting as a principle on which to examine place as elemental and converging realms of importance and meaning. Knowledge of how living things fit is not just a body of information, but is a system of action, interaction, and connection (Rose 2008b). Wimberley (2009) describes this using the language of interrelating realms. These are the realms of the personal, social, environmental, cosmic and spiritual. Quite plainly he proposes that, "changes in one portion of the system can be felt throughout other systems" (Wimberley 2009: 7). An interlocking that holds elements in relation, is what creates a greater presence, which is the place world of an ethnic and cultural group. Within this place world are found ways to ensure effective communication between and among elements. Communication breaks down as elements are erased or when new elements are introduced. Maintaining the commitment to decolonising methodologies, and in returning this book to its original methodology of Indigenous pedagogies of place, I wish to explore this notion of elemental presence and absence further through a non-Western discourse.

It is kincentric ecology, an expression coined by Salmón (2000), a member of the Rara'muri Indigenous group of northern Mexico, which defines Indigenous epistemologies of place. According to a kincentric approach, "the world is not one of wonder, but rather familiarity" (Salmón 2000:1329) and "[l]ife in any environment is viable only when humans view their surroundings as kin; that their mutual roles are essential for survival" (Salmón

2000: 1332). "If one aspect of the lasso [which holds together the conditions for kinship and communicating] is removed, the integrity of the circle is threatened and all other aspects are weakened (Salmón 2000: 1329). Rose (2014: 431) also writes of the poetics of 'fit' in a kincentric ecology, taken to mean the manner in which "human culture" itself pulses and flows within the patterns of surrounding life. Those ecologies that fit are imagined as ones in which humans are situated within the communicative world of pulse and flow, and thus within a myriad passionate calls for response and connection (Rose 2014: 441). Rose (2014) illustrates this through an ethnographic account of the communicative pathways between desert, water, rain, Indigenous ancestral narratives and songlines, sacred sites, bird life, smells and the colours of the riverine environments, to name but a few of the interacting elements in the place that is the Simpson Desert of Central Australia.

The methods by which researchers might come to witness and appreciate this nesting of elements in place are necessarily broad and, according to Rose (2014: 432), call into our field of focus, ecology, ethology, biology, geology and other 'natural' sciences along with social and cultural ethnography. Bearing in mind the multiplicities that converge on place, and the potential to realise these as outsiders and to witness them through the testimonies of others, then it would seem that disregarding the effects of violence and harm on place itself, whether directed at the ecology, biology, geology or the social and cultural lives of those who inhabit place, is not an option. Kincentric ecology, hyper-relationality and relational ontology combine to reveal the communicative pathways that may exist between people and place elements. While common in some cultural contexts, and reinforced epistemologically by an Indigenous pedagogy of place, these principles of relatedness are often intrinsically denied by others. This is the result of axiological adjustments that have come with particular ways of knowing and being in the place world. The proceeding discussion of place harm, as instituted by toxicity, loss of non-human life, and place death, takes kincentric ecology and relationality between people and place elements as its guiding light, while drawing attention to the permissibility of harm that comes with the denial of kinship and axiological worth in place.

Toxicity

> When one's land is befouled, then, it is like an injury to the family, a wound to the self.
>
> (Erikson 2011: 44)

When place relations are prefaced on kinship, elemental change will exact costs, including what Lynch (1972: 190) calls psychological ones; referring to the "disorientation, fear, regret, rage, or desolation that change may bring". "When an entire landscape has 'gone to waste' we face one of the most painful environmental changes, and one moreover that we are least prepared to deal with: shrinkage and abandonment" (Lynch 1972: 191–192). The human cost is great, and in the face of this, we may "shut our eyes to the suffering or try to prevent the change by subsidies and exhortations" (Lynch 1972: 192). Shifting the focus momentarily away from the human struggle amid landscapes gone to waste, this discussion moves to consider the experience from the perspective of place. In the register of harm, what if violence is directed explicitly at place? And does this matter only once it has begun to infect aspects of human life?

When place, and its elemental parts are wounded, this in and of itself is cause for concern, not only because of the eventual threat it may pose to human life or the security of the ethnic group, but because place matters. The elemental changes that come from poisoning, toxicity and contamination, enacted by invading others or those without care or concern for kinship in place, can only occur if place is stripped of its axiological merit and importance. Industrial development, the construction of major infrastructure projects, and nuclear testing across an ethnic group's homelands are just a few practices that reveal the extent to which others may determine that place does not matter. This has the potential to kill off elements in place or severely compromise their capacity to endure. That the suffering resides not only in the human realm but also in other elemental realms of place does not discount the profundity of loss or the harms felt.

Very few have cast the experience of disaster and toxicity more evocatively than Kai Erikson (1976, 1994, 2011). His account of the destruction of community in the wake of the 1972 Buffalo Creek floods in West Virginia, caused by the bursting of a coal slurry impoundment dam, along with his reflections on the effects of the 1954 US atomic testing on the people of Bikini and Utrik atolls in the Marshall Islands, are riveting and saddening. The human toll is huge, the psychological one accounted for in his work even greater. The psychological effect of contamination, toxicity and destruction on residents and Indigenous owners, in both cases of 'disaster', is traced to the catastrophic elemental erasure that was witnessed and experienced directly in and by place. Wretched from beneath people's feet, hidden by layers of sludge and debris, burnt to ashes, left to rot, oppressed by the weight of radioactivity, place suffered awfully as a result of these events. Perhaps most confrontingly, both events were the chaotic result of human indifference to place value and a failure to imagine place as a sentient co-presence. Blind disregard and denial of kinship rendered possible a sequence of events in which people and place were utterly wrecked by collapse. Returning to Erikson's 1976 text, *Everything in its Path* and his 2011 reflection, 'The Day the World Turned Red' I am struck by the images that the following accounts inspire:

> [A] burning black wave lashing down the hollow and taking everything in its path. The ears can almost hear a roar like thunder, pierced by screams

and explosions and the crack of breaking timbers. The nostrils can almost smell the searing stench of mine wastes and the sour odor of smoke and death and decay.

(Erikson 1976: 186)

On 6:45 on the morning of 1 March 1954, a thermonuclear device code-named 'Bravo' was detonated seven feet above the surface of one of the islands that made up Bikini atoll. The island simply disappeared.

(Erikson 2011: 27–28)

As I reread these accounts, I am drawn to wonder on the experiences of place and its elemental parts. To all those co-presences, for which the sociological imagination has had less concern, this discussion of toxicity is committed to witnessing. These are the non-human species of flora and fauna, the geological communities and organisms, the ancestors, and when all combined, the universe that culminates in place. There is no doubt that each and all suffered and the overall well-being of place was severely impacted by their decline.

The hazardous effects of exposures to toxic chemicals, radiation and biological and physical agents are evident throughout the place world, encountered in spaces where conflict and demand (as modelled by extractive relationships and Western 'developmental' principles) outweigh relations of mutuality and co-dependency. Recently, Povinelli (2014) has tackled the presence and principle of toxicity. Opening up a rich and timely dialogue, she writes: "Toxicity figures in a range of contemporary political, economic, social, and environmental discourses, from the toxic waste of the gulf catastrophe or Fukushima and the toxic assets of financial institutions, to concerns over toxic lifestyles and the biomonitoring of toxic bodies" (Povinelli 2014). The theme of toxicity is invoked and explored through improvisational realism in the film *Windjarrameru, The Stealing C*nt$*, produced by the Karrabing Film Collective, a grassroots Australian Indigenous based media group, of which Povinelli is a member and performed the role of director. It tells the story of a group of young Indigenous men hiding in a chemically contaminated swamp after being falsely accused of stealing two cartons of beer, while all around them miners are wrecking and polluting their land (Povinelli in correspondence dated 15 September 2015; Karrabing 2015). In the midst of this unrestrained assault on the place world, sacred areas "are being decomposed for their mineral content and recomposed as composites of crushed beer cans, plastics, wire, and chemical effluents" (Povinelli in correspondence dated 15 September 2015).

The film and Povinelli's commentary on other related matters of toxicity and contamination across neighbouring regions compel audiences to reflect on why the story of toxicity matters. The search for answers to this provocation, however, requires asking the deeper question, whose lives or what lives matter? (Povinelli in correspondence dated 15 September 2015). In the fullest realisation of what does matter, the harmful effects of toxicity and poisoning on ancestral co-presences and Dreaming beings as the very substance of place are brought to light through the Karrabing project. The multifarious agents that reside in place, as inclusive of human and non-human presences, are seen to live and struggle together through contamination. When faced with toxic contamination, the ontological order of human life is intruded upon, but for the Karrabing, and other Indigenous communities, an equally pressing concern is how the Dreaming and ancestral beings that are place itself might mutate or change as a result of chemical mess and disorder. Rendering this in specific and regional terms, Povinelli asks, how might the health and welfare of the mermaids who swim through the aquifer tunnels of a landscape be affected? That as contamination and toxicity set in, people and place may simultaneously decline is an ontological certainty according to Indigenous epistemologies. Thus, asking what are the levels of toxicity affecting the average Indigenous person since these disordering structures were put in place, is equal to asking, what is the health of place at this time (Povinelli in correspondence dated 15 September 2015).

Because place is an "ecological condition" (Povinelli in correspondence dated 15 September 2015), not just a spatial field, and because it is kinship obligation that scripts the human role in maintaining this condition, the very question of toxicity's effects must canvas both the place and human cost exacted through dissociative relations that lead to chemical contamination. Toxicity is a serious agent of place harm, and how it comes to be a weapon for those seeking to culturally wound in times of ethnic conflict, or to rescript place as extractive quarry for human benefit, is traceable to the logic: 'give human life little choice but to die or leave'. Contaminating and insidious silent toxins "scare human beings in new and special ways, … [and] elicit an uncanny fear in us" (Erikson 1994: 144). As a result of such encounters, people and place are no longer perceived as orderly, there is no narrative structure to a toxifying event or slow poisoning. Toxic disasters "violate all the rules of plot" generating epistemological confusion and ontological uncertainty (Erikson 1994:147). As Peeples (2011: 373) reflects: "Imaging toxicity is no simple task as many pollutants are invisible and sites of contamination are concealed, especially for those of privilege." Through this hidden dimension, and its concealed effect, toxicity is a means of achieving violence against place that is often severely punishing for its human residents and kin.

There is an insidious creeping nature to toxicity, evidenced by uncertainty as to the extent of harm, or the slow decline of floral and faunal species with catastrophic losses revealed only in decades to come through birth defects, or cancer and other disease epidemics. Nixon (2011) writes of toxicity as one part of a broader and slow-marching violence of environmental decay, inclusive of climate change, deforestation, oil spills and the environmental aftermath of war. He draws attention to the attritional lethality of such events. The violence is slow because it is so readily ignored by those empowered to control its arrival and acceleration. It is a hard-charging capitalism, which is held to exacerbate the vulnerability of ecosystems and of those who are structurally, economically and politically marginalised (Nixon 2011). These people are the poor, the disempowered and often involuntarily displaced. They become targets in many instances because of ethnic difference or perceived undesirability as a collective presence in place.

Toxicity and the slow violence of environmental decay stems from and also fuels social conflicts that arise from desperation as life-sustaining conditions erode (Nixon 2011). Writ-large, toxicity brings health and environmental risks and induces strain through the erosion of certainty in place. The psychological effects of this are profound. So too, "[f]or those living near a polluting industry, the detection of contamination often occurs when the toxins manifest themselves in the body through pain or disease" (Peeples 2011: 374). As if facilitated by the bonds of kinship, the effects of toxicity on place come to pulse through the human body over time, with the rhythms of toxicity following the principle of nesting, whereby otherness is always in relation, and the effects on one element in place reverberate through all others.

Whether seen or unseen, toxicity disorders place, rupturing its position within a cultural geography, by inducing conflicting imaginings and realities of place. Places harmed by toxicity are propelled between states of beauty and ugliness, magnitude and insignificance, the known and the unknown, inhabitation and desolation, security and risk (Peeples 2011: 374). They are simultaneously represented as wasteland (landscapes ravaged and made dangerous by technologies of the modern era) and wonderland (remote landscapes conceptualised as deserted, and quietly traumatised) (Brown 2013: 27). It is from the unnaturally coloured tailings ponds of a mining site, the eerie, emerald luminescence found in poisoned waterways, kilometres of desiccated landscape and dead forests and the smell of pollutants from burning industrial waste that the revelation comes, 'something is wrong'. By the time this realisation occurs, significant harms may have already been done, and surely they have already begun in place. This induces trauma among human populations for the futility in trying to stop or contain toxicity, and comprehend the extent of its spread. Forever altered by "human technology, weaponry, and industry", many of the most contaminated sites on earth were and are places that matter, but which have since been destroyed, disordered and designified. Coming back from such punishing sentences is a gruelling exercise for place and its people.

Deliberate poisonings, killings and ecological decline are all part of the weaponry of cultural and environmental wounding. They conspire to undermine human rights and place ethics, often working to precipitate the death of ethnic groups and entire communities (see Hinton 2002; Totten et al. 2002). One such example is the poisoning of the Ok Tedi River and the Fly River in the Western Province of Papua New Guinea, by BHP Billiton mining operations. Illustrative of violence in place, this poisoning was made possible by a combination of denying kinship in place, axiological retreat and moral disengagement with place, on the behalf of the hard-charging capitalism of the mining industry. It resulted in an assault on place elements, comprehensive enough as to bring toxicity and destroy life. Kirsch (2007: 305) reports that the original environmental impact assessment called for the construction of a tailings dam in the mountains. A landslide at the construction site led the government to temporarily permit the discharge of tailings and other mine wastes directly into the Ok Tedi River (see also Hyndman 1988; Townsend 1988, cited in Kirsch 2007: 305). Protests against the Ok Tedi mine began in the late 1980s as the first signs of toxicity began to appear; the Ok Tedi River was polluted and gardens along the river floodplain showed early signs of destruction (Kirsch 2007: 305).

Regarded an 'environmental (human induced) disaster' without parallel, more than 1 billion metric tonnes of tailings (mining waste) were discharged into the waterways from the mid-1980s (Kirsch 2007: 305; see Low and Gleeson 1998 for a discussion of this case relative to environmental justice discourse). While the mining company did not explicitly undertake a programme of regional poisoning, it is in the axiological crisis of disregard for place worth and the dismissal of Indigenous rights and place value that ensured the land-use practices that led to the poisoning. Indigenous people throughout the region have expressed anxieties over contamination for more than 40 years, yet not necessarily calling for the mine's closure and end to operations. There is deep complexity in how these operations are negotiated by Indigenous groups, and Kirsch (2007: 314) notes that people continue to grapple with having to choose "between environmental degradation and monetary compensation" for damages done. When economically and politically marginalised, Indigenous decision-making operates across difficult terrain. Kirsch illustrates the paradox faced by Indigenous owners of lands and waters affected by the mine:

> they want to simultaneously protect the environment and have access to development opportunities and money. It is almost impossible for them to contemplate early mine closure when so much of the regional economy is dependent on its operation, and when so many of the villages have already been significantly affected by pollution and need the extra income provided by compensation payments to feed their families.
>
> (Kirsch 2007: 314)

There would appear to be form of unmitigated cruelty in the mine's very existence and the conditions it has created, thus rendering local decision-making part of a deeply wounded scape of coloniality and neoliberalism. These decisions are made against a backdrop in which forests have died and remain under considerable stress, thick grey sludge has raised the riverbed and disrupted Indigenous transportation routes, and fish counts have declined. Local Indigenous owners of the land and waterways have consumed contaminated fish and contamination has leached deep into the mud and the region's staple foods. "Looming on the horizon is a new environmental threat: acid mine drainage, in which sulfuric acid leaches heavy metals into the river system, rendering the affected areas inhospitable to organic life for decades. Reports from the mining company indicate that acid formation is already occurring at low levels and will increase in the future" (see OTML 2005: 1–2, cited in Kirsch 2007: 314). The violence of this event cannot be understated, and raises questions as to the meaning of death and decline and accountability in regards to all forms of death (Rose 2013). Toxicity that causes entire sago forests to die is surely a deep violence against place, akin to the poisoning of human life in a nested ecology.

Totten, Parsons and Hitchcock (2002: 71), writing on the pervasive and harmful effects of 'development projects' led by multinational companies, such as BHP Billiton, consider the experiences of Indigenous people worldwide, examining the forms of violence they and their homelands have endured. Ecuador, West Papua, Burma, Nigeria, Tanzania, Malaysia and Australia all figure in the depressing ledger of recurring harms, including forced relocation, massive environmental problems, oil spills, water poisoning, loss of biodiversity, slavery, murder, habitat destruction, deforestation and destruction of sacred sites. Like a band of awful brothers, these methods of violence against people and place occupy the same space, simultaneously wrecking their harm against place and people as co-presences that present obstacles to invading agents. Often rationalised by arguments of inevitability, accidental consequence or overcomeable obstacles, the decline of one is implicated in the decline of the other. That all of this has been done in the name of 'economic progress' in cases of mining and large-scale infrastructure development does nothing to lessen the effects of violence on place, nor toxicity that lingers chemically, psychologically and socially. "The poor treatment of indigenous peoples and the loss of their land has had a series of negative effects, including reduction of their subsistence base, nutritional deprivation, and heightened social tensions, some of which are manifested in higher rates of suicide" (Totten et al. 2002: 73).

Whether toxicity, contamination and polluting of place can be deemed anything other than purposeful and intent-laden is a contentious point worth raising here. Knowledge of potential risk and harm is ever-present in large-scale extractive industry settings and in the economic progress of infrastructure and resource development. It is epistemologically shrouded in the rhetoric of development for greater good, a logic that is traceable to certain ways of knowing and being in the world (see Grosfoguel 2007 for an enlightened critic of political-economy paradigms). Through separatist principles of nature/culture and the primacy of the human in an ecological supremacy of development, a hierarchy is constructed in which place value, as ecologically, culturally or environmentally conceived, is relegated to the lower rung.

A great number of toxifying episodes in place, from mining to nuclear testing, are made possible by this relinquishing of place to background status. When veiled by economic rationalism, or developmental rhetoric, they eventuate because of a moral disengagement with the ethics of care. The by-product of epistemologies and ontologies that shape place as non-place, this moral disengagement reaches one of its highest orders in the case of nuclear testing (see Davis 2005a). A very particular form of violence, absolute in its capacity to harm, it induces pervasive and lasting forms of toxicity. As a practice, nuclear testing has been historically justified by the need for 'new and additional information' pertaining to military arsenal and its effects. "[B]etween 1945 and 2006, over 60 locations throughout the world were used to detonate more than 2,000 nuclear explosions for military or for peaceful purposes" (CTBTO nd). Describing the place world of nuclear testing, the Commission for the Comprehensive Nuclear-Test-Ban Treaty Organization (CTBTO) describes, "idyllic South Pacific atolls", "Novaya Zemlya … a remote ice-bound archipelago in the Arctic Ocean", Lop Nur "a landlocked salt lake marsh" in China, and the "arid desert test site of southeastern Xinjiang Uygur Autonomous Region", home "only to roaming wild camels".

These regions had varying demographics at the time of testing, some densely inhabited with Indigenous residents, others not so, habited mainly by non-human presences. With or without human presence, there is consequence; as the ecology and place elements of Pacific atolls, ice-bound archipelagos, salt lake marshes and arid deserts are fully attested to in the realm of natural science, through geological, geographical and ecological inventories. The sentiency of place if not dependent on human presence, is found in such locations as part of the nested ecology that contains multifarious life forms. Many of these former test sites remain testing grounds today, with the focus having now shifted to the harmful and enduring effects of radioactive contamination, ecological decline and remediation.

The enduring effects of contamination are such that places where nuclear testing has occurred have either disappeared, leaving a gaping hole in the place world, or been deemed unsuitable for sustaining life in any way. The following account of violence and trauma in place presented as part of this diagnostic discussion engages these very concerns. It is focused on the United States nuclear testing programme in the Pacific Region of the Marshall Islands. The Marshall Islands consist of 29 atolls and five individual islands in the Central Pacific (Maragos 2011:123). During the years of the Cold War, from 1946 to 1958, 67 nuclear weapons were tested across the 'Pacific Proving Grounds', with the largest US nuclear test, code named Castle Bravo taking place in 1954 (Davis 2005a: 607; Guyer 2001: 1372).

Toxicity and exile in the Marshall Islands

The Marshall Islands are home to the Marshallese people. They are the Indigenous owners and hold customary land tenure, through matrilineal descent, over the expansive islands, atolls and reefs of this Pacific region. Clan boundaries are distinct, and ancestral beings are present in and across the land and sea (Barker 2004: 11–14). There are sacred sites across the vast network of islands, where both plant and animal life are highly revered and linked to the domains of senior ancestral beings (Barker 2004: 14). The entire Marshall Island region is defined by Indigenous epistemologies of place and kincentric ecology (see Barker 2004; Tobin 2002). In the wake and aftermath of nuclear testing, the atolls, reefs and islands were each transformed in some way. In the phase leading up to testing, place transformation was the subject of US military imaginings. The islands and atolls were cast as desolate 'non-places' (Davis 2005a). This ensured the second phase of place transformation, which resignified the Marshall Island complex as sites of contamination, relocation and remediation.

The nuclear testing that was henceforth carried out is regarded, in the context of this book, as an act of violence directed at the well-being of the Marshallese Islanders, prefaced on a disregard for human life and place value. The US government conducted nuclear testing with knowledge of potential risk and harm, and instated policies of relocation that were certain to bring about cultural wounding and hardship for the islanders. While not in the custom of war or genocide as elsewhere documented, the dismissal of human rights, the intent to induce harm and suffering both in people and place is a form of ethnic violence. The people and places of the Marshall Islands had to be designified of meaning, and resignified as non-people (or quasi-human) and non-place (or place as mere backdrop to human life) in order for events to transpire (see Davis 2005a).

That the reinscription of people and place as 'testing ground' beyond an ethics of care was achieved is conveyed through the testimony of an official of the Atomic Energy Commission, the agency that had conducted the tests. In a statement on the exposed islands it is held:

> [this island] is by far the most contaminated place on earth, and it will be interesting to go back and get good data ... Data of this type has never been available. While it is true that these people do not live the way Westerners do – civilized people – it is nevertheless also true that they are more like us than mice.
>
> (cited in Erikson 2011: 29)

The effects of nuclear testing across the Marshall Islands have been heavily documented with specific reference to Bikini Atoll as a site of detonation and Utrik Atoll as a neighbouring atoll heavily affected by the raining debris of nuclear fallout. Coercive relocations occurred for residents of both atolls, and also Enewetak Atoll, while and contaminants also heavily affected Marshallese Islanders on Rongelap Atoll (see Johnstone and Barker 2008; Takada 2000). So extensive was the fallout that the nuclear contaminated debris left on Enewetak Atoll remains a significant cause for concern amid growing fears of sea-level rise and the threat of contaminants seeping into the ocean (Bridges and McClatchey 2009; Tabucanon 2014). The Marshall Islands are now the highest ranked nation most threatened by flooding and endangerment due to climate change (Tabucanon 2014).

Specific concern is directed at the Runit Dome, locally referred to as 'the Tomb', on Enewetak Atoll. Constructed in 1979, the large concrete dome houses 111,000 cubic yards of radioactive debris left behind after 12 years of nuclear testing. The contents of the dome are derived from the scrapping off of Enewetak's contaminated topsoil, which was mixed with radioactive debris (Jose et al. 2015). The resulting radioactive slurry was then dumped in an unlined 350-foot crater on Runit Island's northern tip, and sealed under 358 concrete panels (Jose et al. 2015). "Brackish water pools around the edge of the dome, where sections of concrete have started to crack away. Underground, radioactive waste has already started to leach out of the crater: [and] according to a 2013 report by the US Department of Energy, soil around the dome is already more contaminated than its contents" (Jose et al. 2015). Let us now turn to the sequence of events that has made all of this possible.

Bravo, a dry fuel hydrogen bomb was detonated over Bikini Atoll at 6:45am on 1 March 1954. Radioactive fallout from the detonation spread across Rongelap, Rongerik, Enewetak and Utrik atolls. Traces of radioactive material were also found as far as Australia, India, Japan, the United States and Europe (DeGroot 2004:198). When Bravo was detonated, it formed a huge fireball, gouging a crater more than a kilometre wide and 60 metres deep (Blades and Siracusa 2014: 60). A cloud, born of the fireball, smoke and debris lifted nearly 14 kilometres into the sky within a minute (Blades and Siracusa 2014: 60) "As a result of the blast, the cloud contaminated more than seven thousand square miles of the surrounding Pacific Ocean" (Titus 2001: 47). With the detonation of Bravo, "[p]articles of coral pulverized to an almost weightless powder and alive with radioactivity – drifted eastward" (Erikson 2011: 29). A radioactive wave washed over Bikini, causing complete removal of all vegetation, "killing off all animals life except one hardy variety of rats [sic]" (Trumbull 1982: 49, cited in Davis 2005a: 616). The atoll was littered with testing equipment and damaged facilities, and most of the coconut and pandanus palms were gone (Davis 2005a: 616). Radioactivity entered the atmosphere, the water, and the soil and worked its way up both terrestrial and oceanic food chains (Guyer 2001: 1373). Within a week of the Bravo disaster, high levels of radioactivity were detected in cattle thyroids (Guyer 2001: 1373). The soil and water contained radioactive isotopes of caesium, cobalt, strontium, americium and plutonium, and while the radioactivity in the lagoon and sea dissipated through slow and steady dilution in the ocean, radioactivity on the islands stayed (Guyer 2001: 1373).

At the time Bikini Atoll was home to approximately 150 people, Utrik to 159 and Rongerlap to 64 Marshallese Islanders (Erikson 2011: 29; Davis 2005a: 613). The first relocation of Marshallese began with the Bikinians, who were moved to Rongerik Atoll in 1946 ahead of the commencement of a lengthy period of nuclear testing. Reflecting on events, Davis (2005a: 614) remarks that, "the imagined cultural inferiority of the Bikinians and their supposed weak attachment to Bikini were used by the US military as legitimization for their removal from their atoll". Forced to relocate to

Rongerik Atoll, the Bikinians were separated from their place kin, a home rich in resources, "the site of their lived experience" and the "cultural landscape saturated with meanings, deities, and the graves of their ancestors" (Davis 2005a: 615; Niedenthal 1997). The place in which they found themselves resettled was one deemed "a form of hell where a demon named Litobora left the fish, coconuts, and pandanus poisoned" (Davis 2005a: 615). Two months after the Bikinian arrival on Rongerik, water and food supplies were found to be inadequate and the people began requesting repatriation to Bikini (DeGroot 2004: 330; Davis 2005a: 615). Their requests were denied, and in time people were found to have reached states of starvation and were found to be suffering from ciguatera poisoning (Davis 2005a: 615; DeGroot 2004: 330). Rongerik conveyed an ethos of kinship denial. It was a place that struggled under the weight of an enduring human presence, its order otherwise structured. In 1949, Bikinians were relocated once again, this time to Kili Island, regarded by many Bikinians as a less than ideal location, yet one where many continue to reside today.

As for the fate of the Bikini Atoll itself, in the wake of a series of detonations, which culminated with Bravo in 1954 and persisted until 1958, it is recalled as follows; to begin, the bomb Bravo vaporised two islands of Bikini Atoll and part of Nam, the island at which it was detonated (Guyer 2001: 1372). Brown (2013: 31), through a series of archaeological surveys, notes that "the physical evidence of nuclear testing at Bikini Atoll comprises extensive landscape and seascape modification", including the crater caused by Bravo (two kilometres wide, 80 metres deep) and other detonations, 21 experimental vessels that were sunk into the lagoon, the remains of bunkers and other built structures. Adding to the cruelty of this venture, throughout the history of testing in the Bikini region, and in particular during the Baker testing in 1946, several of the experimental vessels that were sunk had guinea pigs, mice, pigs, goats, rats and grain containing insects placed on deck. They were also placed on land, to simulate human responses to nuclear blasts and fallout (Bonner 1997: 13; Winkler 1999: 90). All succumbed to the effects of the blasts and nuclear fallout, or were killed shortly after by human intervention (Niedenthal interview cited in Bartholet 1997; Navy History and Heritage Command 2002). Etched with tangible signatures of the event, Brown (2013) cautions that more pervasive is the heritage that is etched radioactively into every particle of the land and seas of the atoll.

In the late 1960s, people were 'returned' by US Military forces to the "renovated paradise" of Bikini Atoll, replanted with coconut palms and somewhat cleaned up of debris (Davis 2005a: 616; DeGroot 2004: 330). Most Bikinians refused to return amid fears of contamination and impending sickness. Some returned in the 1970s, only to be expelled some five years later, with growing concerns over exceedingly high levels of radioactive contamination in the bodies of atoll residents (Davis 2005a: 616). Tests revealed that the island would remain uninhabitable for at least a century (DeGroot 2004: 330). According to Davis (2005a: 617), "in the minds of many Bikinians and non-Bikinians [the

view is] that Bikini is a contaminated place that ought to remain deserted".
Today, while people consider it safe to venture for short periods to the atoll,
longer stays and the consumption of local foods is considered too risky a venture (Davis 2005a: 617).[2] Bikinians continue to live in exile from their home
and the atoll continues to harbour the invisible contaminants of the violence
enacted against it.

In the wake of Bravo, the residents on neighbouring Utrik and Rongerlap
atolls were also relocated, but only after the threat of fallout eventuated.
Many experienced immediate radiation sickness; others developed serious,
long-lasting, and ultimately fatal illnesses months and years after the blast
(Guyer 2001: 1372). After three months they returned home to Utrik, to
find a new world. Their chickens and pigs had all been slaughtered by the
Americans, they were instructed to not eat local food or drink local water, "no
coconuts, arrowroot or fish from the lagoon, no breadfruit or taro – nothing
that grew on that strip of land" (Erikson 2011: 33). To instruct them in such a
task, was, according to Erikson (2011: 33), "like asking them not to breathe",
thus can be the nexus between people and place.

While the human death toll, from the immediate fallout of Castle Bravo
was greatly reduced by displacement, there was death of another kind. Lore
Kessibuki, leader of the Bikinian people at the time, is recorded as saying: "Bikini is like a relative to us" (DeGroot 2004: 330). "Like a father or a
mother or a sister or a brother, perhaps most like a child conceived from our
own flesh and blood. And then, to us, that child was gone, buried and dead"
(DeGroot 2004: 330). Destruction and toxicity have become part of the bedrock of place, and the right of the Bikinians and other Marshallese Islanders
"to live in a safe environment and in their own land … was trumped by shows
of military might at Bikini Atoll" (Guyer 2001: 1375). Both at that time and
in the intervening years, other rights, including "the right to protection from
harm, the right to have their autonomy respected, the right to be told the
truth, the right to just treatment – also have regularly been ignored" (Guyer
2001: 1375). Davis (2005a: 613, 614) asks a very important question of these
events, namely, how was it that Bikini, an inhabited island, became the site
for these tests? She finds her answer is twofold and traced to the ability, on
the one hand, to render this place a non-place (a blemish on an otherwise
vast unpopulated ocean) and on the other to represent it as 'backward', the
antithesis of America's technological modernity. Both involve a stripping of
meaning and the denial of kinship in place, necessary steps to allow for the
bombing and toxification that was forthcoming.

Counter to such views, the Marshall Islands make up part of a sentient
land and seascape that is populated by coral reefs, marine mammals and ecologies, along with human kin and ancestral presences. All have been affected
by the nuclear testing that has historically 'taken place'. While there remains
some debate as to the effectiveness of remediation efforts across the islands
and atolls affected by the testing, and the presence of harmful levels of radioactivity, the order of place and its character has been fundamentally and

irreversibly changed. Irrespective of whether radioactivity does or does not continue to post a threat to human life, the lingering effects of 'perceived risk' remain real and harmful (see Davis 2005b). Questions of whether a place is ever 'really safe' in the wake of toxic episodes pervade the psyche and shift the nature of kinship between people and place.

If place is perceived to contain or harbour threatening substances (visible and invisible), this is often suffice to shift its conceptualisation in the hearts and minds of residents and visitors. This can lead to emotional geographies of fear, grief, reticence and disgust, all a far cry from functioning kinship in place, which leads to affirming experiences for people and place alike. Davis (2005b: 213) reflects that "scientific studies of risk operate as cultural systems", thus cross-culturally the risk of misunderstanding the effects of toxicity and poisoning is ever present. In cases where people may not have the ability to avoid a 'dangerous place' (Davis 2005a: 213), whether as the consequence of having nowhere else to go, or due to the emotional connections and obligations born of kincentric ecology, a profound grief may emerge from witnessing the decline and suffering of place or the sense that place kin is now capable of inflicting harm upon them.

According to Davis (2005b), in instances of toxicity, deciphering the status of place, as 'safe' or 'unsafe', is achieved through two primary modes, a 'realistic perspective' or 'constructivist perspective'. Leaving aside the obvious critique begged of what constitutes 'realism' in this case, these approaches to translating place harm and impending risk, in the first instance, assume risk is "[a] tangible by-product of actually occurring natural and social processes. It can be mapped and measured by knowledgeable experts, and, within limits, controlled" (Jasanoff 1999: 137, cited in Davis 2005b: 213). In the case of a constructivist approach, risks "are refracted". They are passed through "lenses shaped by history, politics and culture" (Jasanoff 1999: 137, cited in Davis 2005b: 213). Both are cultural perceptions (Davis 2005b: 214; Beck 1999: 135). It is the latter, however, that aligns most closely with kincentric ecology and it is the former that instates the greatest distance between place and people (as envisioned through a rhetoric of nature as a distinct and separate realm from culture). Illustrated by the Bikini Atoll experience, while "most agree that Bikini Atoll was contaminated from nuclear testing and that exposure to radioactivity has had deleterious effects on people and other forms of life", what is now contested "is the extent to which Bikini is a dangerous place today" (Davis 2005b: 216). The investment in deciphering this one way or the other is markedly different for Bikinians, other Marshall Islanders and visiting Western scientists. In the end, for those who are kin to this place, it remains all 'muddled up' through uncertainty of rapid change and the enduring possibility of harm that toxicity emplaces. It is clear that toxicity, actual or otherwise, has abiding and long-term effects. What better term to convey this than 'root shock'; as an enduring harm that resides deep within the corpus of place.

As discussed in Chapter 1, root shock seeps into the bones of people, the sediment and axiology of place (Fullilove 2004; Erikson 2011: 37). It is because of the value of place, as understood by those for whom it matters, that attempts at fundamental disorder must be driven back by the constant referral to place value and meaning. That said, the story of toxicity is a long one, thus the enduring and exhausting plight of those who live with it. It is punctuated by efforts at resettlement, repeated displacement, remediation, settlement claims, illness, mutation and death. The surveillance of place often continues as the human experience of its harm falls away from sight or, alternatively, place is left to quietly disappear while people are ushered elsewhere into their future, often an uncertain one. Povinelli observes a third possibility, as is conveyed through the film *Windjarrameru, The Stealing C*nt$,* (Karrabing 2015), whereby "Indigenous sovereignty over space fully reemerges in the space of utter state abandonment and total capital despoilment" (Povinelli in correspondence dated 15 September 2015). This is described as "the grinding contagion of settler colonialism", and raises the possibility that Indigenous sovereignty may thrive once again "where Europeans have come, destroyed, and are [now] fearful of returning but to which the Karrabing [Indigenous owners] continue to hold on" (Povinelli in correspondence dated 15 September 2015). The ability to hold on and prevail amid conditions of total and absolute contamination is directly attributable to the very fact and pervasiveness of kinship. There may be no option to leave, just as there is no option to leave one's most beloved of family members alone and in harm's way.

How the experience of contamination and toxicity is made sense of is difficult to explain, nor can it be simplified down to a sequence of responses and adjustments that might make sense to all. This fact is so obviously conveyed by Goodall in her 1994 work on colonialism and catastrophe. Goodall (1994) documents memories of nuclear testing on Indigenous homelands in central Australia. In this case, between the years 1957 and 1963, the United Kingdom conducted nuclear tests in mainland South Australia at Maralinga and Emu field, passing a low cloud of radioactive smoke and particles northwest over Indigenous homelands located in and around the small towns of Mintabie, Wallatina and Welbourne Hill (Goodall 1994: 55). Despite an official campaign of British forgetting, memory of the event has endured for the Indigenous people who saw this cloud, lived with its effects or now recount its mournful presence as oral history.

The Yankunytjatjara, Anangu and Pitjantjatjara people became the first mainland Aboriginal groups to face contamination from nuclear weapons (Goodall 1994: 57). Goodall worked as a researcher tasked with drawing together accounts from available documents and from Aboriginal and non-Aboriginal memories for the purpose of a Royal Commission into British nuclear tests in Australia (Goodall 1994: 58). Oral testimonies revealed that people remembered the "fearful experience of a low drifting black mist stretching as far as they could see, passing over their camps, leaving deposits on the ground and resulting in immediate and longer-term illness" (Goodall 1994: 57). British and Australian governments attempted to clean the site three times: in 1967 (which left plutonium and other radioactive contamination), in

2000 and again in 2009. There are claims, however, that contamination still plagues the site (Leschine 2014: 43; Creative Spirits nd).

The nuclear testing of 1953 was bracketed temporally (1948 and 1957) by two measles epidemics that severely impacted Indigenous populations in the region. Goodall (1994: 60) reports that the Anangu were unable definitively to distinguish their memories of illness caused by nuclear testing from those of illness caused by measles and other infectious diseases. Documents pertaining to the 1948 measles outbreak suggest that between one-quarter and one-third of the Anangu population at the Ernabella mission and on smaller properties throughout the region died within a couple of weeks (Goodall 1994: 63). Nine years later, a second measles epidemic hit, affecting mainly those children under nine who had grown up without the immunity acquired by the survivors of 1948. Epidemics play a major role in shifting the balance of power between groups in interethnic encounters. Where cultures have collided on biological and geographical frontiers, the spread of diseases such as measles, smallpox and influenza has marked the medical histories and the narratives of many Indigenous groups. The loss of life as critical mass and presence of an ethnic group, coupled with the loss of political structure and the transmission of powerful knowledge from elders to younger generations, has a profound effect on all expressions of identity (Kearney 2014: 54–55). This creates particular relations on the enduring frontier and even conditions of fear amid often-desperate efforts to make sense of these illnesses and seek remedies for them from the existing suite of medicinal knowledge. Recorded oral testimonies reveal that for the Anangu, the measles epidemic became entangled in perceptions of the nuclear testing, as it occurred around the same time as the tests (Goodall 1994: 63).

These disasters were etched into the social memories of Indigenous groups in the region, "for some as closed episodes of pain and grief, and for others as continually troubling wounds that refused to heal because there had been no satisfactory explanation" (Goodall 1994: 63). As simultaneously occurring and toxifying events, Goodall found that intensifying colonialism, disease epidemics and nuclear testing became part of a singular moment in Indigenous memory. As they merged into an overarching form of wounding, distinctions were not necessarily made when tracing cause and effect. These violent encounters coalesced at a moment in time, forming a coalition of harm. They are recalled as a sequence of relative tragedies, each made possible by the denial of value, in both people's lives and place's importance. The dismissal of place importance, which made possible the nuclear testing, is directly aligned with the disregard for Indigenous lives. Thus human life, non-human life and the wider ecology of place were merged into a singular object of axiological disregard by outsider colonial agents. The entwined histories of disease and nuclear fallout reveal the extent to which the human body and its lived experience can suffuse with place. That these events were often indistinguishable, or transposed through memory over time, is not merely about people's inability to recall or separate out moments in time, but as oral testimonies recorded in the 1980s attest, the root shock caused by these events was deeply felt. The relentless search for meaning unites the suffering of people with that of place.

Bearing witness most consciously to this relationality as an ethics of care, this discussion moves now to the final form of place harm addressed here, namely loss of life as species decline and ferality in place. It would be remiss to discount the experiences of non-human kin that present themselves as part of a sentient world. How people cope with rapid and often traumatic loss of non-human kin, and the invasion of species (beyond the human) with no regard for the law or order of place is deeply complicated. It reflects practices of communing between people and place, and may call upon causation theory that involves ancestral spirits or otherly agents and co-presences. It may cause a "complete loss of faith, [or] disillusionment in the traditional belief system[s]" that instil meaning in place and the world (Goodall 1994: 68) or, conversely, be accommodated through sophisticated epistemological structures and intellectual negotiations of new phenomena (Trigger 2008). What is overwhelmingly clear is that in the service of cultural wounding, and guided by the intent to harm, non-human species have also suffered at the hands of violent perpetrators and in the midst of ethnic conflict. Harming the species on which human life depends is no veiled attempt to bring about human suffering and the decline of place and ecology, for it erodes the foundations that support the nested elements of life, as consisting of the personal, social, environmental and spiritual.

Loss of life: deliberate killing, species decline and ferality

Whether the widespread killing of bison across North America by marauding settlers, or the poisoning and shooting of dingoes by settlers and farmers in Australia, there is much evidence to suggest that the killing of non-human animal life is a weapon of destruction and outward expression of ethnic hate and disregard for existing economies and cultures. Not only does human conflict, in the form of war, destruction and ethnic conflict lead to the deliberate killing of non-human animals, it also creates ecological dead zones, instating conditions that render place incapable of supporting life. Non-human animal populations may suffer due to starvation, disease or because of competition over resources with the arrival of new, invasive and feral species. They may be attacked in an effort to eliminate food resources and transportation available to people under threat, they may be used as delivery systems for explosives (as with records of bombs strapped to donkeys, horses, bats and pigeons), or have their habitats and waterways poisoned in an attempt to starve out and cause the demise of all living presences.

Where there is contest over lands, waters and resources, incoming agents of harm require the instating of what Rose (2004: 61) describes as a 'Year Zero', a practice tantamount to a "declaration of war" on people, place and place elements. This ensures that the social and environmental world is wiped clean of a history, and left open for a new future, one that can be written onto an alleged 'blank canvas' in the place world (Rose 2004: 73). This is a "violent thrust" into the existing space and time of place (Rose 2004: 54) and that there is little to no room for existing place order and place elements is certain. The statistics of non-human animals deaths precipitated by such realities (whether deliberate or 'consequential') are difficult to find, and certainly the frequency with which they are reported pales in comparison to those of human death tolls in times of conflict.

The record of non-human animal casualties in conflict zones that does exist has been heavily skewed towards a discussion of animal presences on the battlefront; in particular the horses, donkeys, dogs and carrier pigeons of wartime (see, for example, Hediger 2013). These are the creatures forced into servitude as messengers, danger detectors and transport facilitators. They have been deployed into direct combat, provided rescue and protection and shared companionship with their human wartime compatriots. But for the most part, how non-human animals, beyond the domesticates of wartime service, experience ethnic conflict and violence in place is secondary to our human imaginings of conflict, and registers of what and who suffers as a result of human aggression. That these deaths are somehow inconsequential seems suggested by the lack of actual detail on non-human animal experiences (particularly of death) in times of ethnic conflict, human violence and war. This jars with the book's methodology, most notably because it diminishes kinship between human and non-human animal species.

Navigation of the unsettling ideological terrain that attitudes of 'inconsequential death' normalise has been helped greatly by the reflections of Matthieu Ricard. Ricard (2013, 2015) considers the matter of 'concern' for the fate of non-human animals. Having written on altruism and the inflicting of suffering on non-human animals by humans, he found the response to his work was as follows:

> the complaint I most often heard during interviews and in TV and radio talk-shows was that my concern about the fate of animals was 'offensive', because so much suffering already afflicts human beings. To my critics, to be concerned for the welfare of animals was an insult to mankind.
>
> (Ricard 2015)

This argument, he contends is "based on the idea that devoting some of our thoughts, words, or actions to reducing [or acknowledging] the unspeakable suffering deliberately inflicted on animals is indecent because it distracts us from the task of alleviating human suffering". Ricard (2015) lays bare the illogic, if not recklessness, of this fallacious position; "[i]f that is the case, then what about the time and energy spent on listening to classical music, practicing sports, or even getting a tan on a beach? Are these activities offensive because they are not preventing the famine in Somalia?"

It may be that we have become somewhat accustomed to reading about violence, murder and political intrigue in conflict zones; "people are murdered, leaders assassinated", yet Villagra (2011: 45) dares us to encounter the discomfort induced by a discussion of non-human animal kinship and murder. She writes of an incident in which members of a gorilla troop were killed. According to reports, seven gorillas, each members of a Virunga troop, known as the Rugendo Family (in Virunga Forest, Democratic Republic of Congo), were slain, murdered in cold blood, and images of their death were circulated throughout international media forums (in particular *National Geographic*, see Jenkins 2008). Reports emerged that these gorillas were not killed for meat, nor were they hunted or poached. Instead their deaths were described as murder, the result of execution-style killings (Jenkins 2008: 35; Villagra 2011: 45). There is strong pull, both in Villagra's writing, and also across media reports of the gorilla deaths, to envision these animals as kin, merciless victims of civil strife in one of the Democratic Republic of Congo's most fraught national parks. In tracing a culprit, the conditions of war, in particular guerrilla warfare between rival militia and the Congolese army, are scouted as causation. It is the thousands of armed soldiers, along with poachers, illegal charcoal producers, refugees and struggling farmers who are described as the potentially threatening co-presences to the Virunga gorilla troops. They find themselves caught within a whole world of human conflict and associated violence. Their survival and/or decline is implicated in the lives of those who now share and equally threaten their place world.

A vexing question indeed. But one that needs asking of all conflict zones, in all times, and all places. The killing of non-human animals and the degradation of the ecosystems on which they depend has proven to be a most successful strategy for harming marginalised populations and those ethnic groups under pressure to disappear or exit place. Illustrated by the cultural injury for Indigenous Australians that has come with the killing of totemic species, in particular dingoes, it is as Rose (2011: 22–23) states, "no death is a mere death ... [t]he death of an animal creates a loss in the fabric of life, a loss that reverberates across other living beings, human and others". Harms done to place and non-human animal kin have been, in some cases, deliberate, as with the aggressive hunting and killing of species to the point or near point of extinction. It has also been the consequence of ecological destruction through land clearing, mining and other unsustainable neoliberal and 'modern' land- and water-use practices. Provocation of decline and decay has also come in the form of vast numbers of introduced and feral species; many that act with disregard, even impunity in a kincentric ecology.

How and why is it that the destruction of non-human animal life can deliver harm to human populations? Why might a declining ecology and loss of non-human species lead to social depression or actual sickness among a group of people? How can this loss trigger a departure from place? When conflict is anthropocentric in its origins and outcomes, why the need for violence that targets other elements of the place world? The simple answer to some of these questions lies in the fact that harm occurs because human life cannot be separated out from the place world, it is nested in and among all other elements. The complex answer lies in the fact that despite Western epistemology's attempts to separate out nature/culture, human/non-human, sentient and rational, into distinct realms of discourse, politics and order, the behaviour that leads to place harm, elemental decay and loss of life, betrays the supressed realisation that, in fact, all is relational. While rhetoric instates a deep separation, violence and aggression reveal an abiding sense and knowledge of relativity and shared co-presencing as pathways along which to deliver ultimate harm. Denial of this remains the cornerstone of modernity. The deliberate killing or consequential death of non-human animal life is designed to undermine the psychological and spiritual balance of people and place, to induce fear, destabilise economies and livelihoods, and precipitate human death or dispersal from place.

Contributions to this consciousness of the greater effects of cultural wounding on non-human life, have come from those who raise fundamental and deeply philosophical questions about the 'animal', sentiency, kinship and suffering. David Abram (1996, 2011), Marc Bekoff (2007, 2010, 2014), Graham Harvey (2005, 2006), Val Plumwood (2002, 2012), Deborah Bird Rose (1992, 2008a, 2011) and Eduardo Viveiros de Castro (1996, 1998) inspire a discourse on animism, interspecies relationships, love, connectivity and entanglement as kinship between human and non-human animals. Far from an exhaustive list of contributors in the fields of environmental humanities, ecological philosophy and cultural ecology, it is specifically their work to which I am drawn in this attempt to better understand elemental erasure and decay in place. What underscores the principle of relatedness across their works is the view that all beings within the place world are autonomous yet connected, "enmeshed in relations of interdependence" while being "self-organizing and sentient" (Rose 2008c: 110). It is this sentiment that returns us to the principle of nesting, reiterating Indigenous place pedagogy. This and the emotional geography that comes with full recognition of co-presencing and kinship is vital to the decolonising of thought on place.

Salmón (2000: 1327) encourages this further by expanding on the parallel principle of kincentric ecology. He cites Laguna Indian author and poet Leslie Marmon Silko, who explains that human beings must maintain a complex relationship with "the surrounding natural world if they hope to survive in [it]" (in Salmón 2000: 1327).

> To Silko, humans could not have "emerged" into this world without the aid of antelope and badger. The Lagunas' sustained living in the arid region of the Southwest could not have been viable without the recognition that humans were "sisters and brothers to the badger, antelope, clay, yucca, and sun". It was not until they reached this recognition that the Laguna people could "emerge".
>
> (Silko 1996, in Salmón 2000: 1327)

In another example of the animism and perspectivism of Indigenous epistemologies, the Rarámuri of northwest Mexico, have a term, *numati*, which describes elements of the natural world as relatives. This is not metaphor, but realism According to Salmón (2000: 1329), in a previous world Rarámuri people were part plant. "When the Rarámuri emerged into this world, many of those plants followed. They live today as humans of a different form. Peyote, datura, maize, morning glory, brazilwood, coyotes, crows, bears, and deer are all humans" (Salmón 2000: 1329). The kinship is not symbolic it is actual and understood as are relations to human kin (Levi 1993, cited in Salmón 2000: 1328–1329).

With these principles in mind, reflecting an Indigenous pedagogy of place, and nested ecology, the effects of ecological and biological colonialism, as forms of violence in and on place and its people can be examined through the lens of place harm and wounding. In the case of colonialism, everything living in the colony, human and non-human, becomes subject to the importation of epistemologies, ontologies and axiologies of separation and ranking (White 2013: 463). The same may be said of any context in which land and waters form the backbone of desires for incoming agents of harm. Whether viewed through the actions of settler colonial nations, nations at war, ethnic groups mobilising to take neighbouring territories, or the claiming of sovereignty over another's lands, wherever a frontier mentality sets in, and the intent is to march forth and instate new order, processes of designification and resignification can violently occur by way of killing and elemental erasure. The frontier pushes ahead over the dead bodies of human and non-human co-presences, 'inconveniences' that are often destroyed in the pursuit of place and power.

New arrivals, as ethnic groups, economies and non-human species, take up where others have once or still exist. The 'inconvenience' of existing order can lead to the deliberate erasure of place elements. Whether directed through deliberate action or through the consequential effects of dramatic change, the result can be a stripping of ecologies vital to supporting non-human animal life. This renders place useless to the ongoing survival of certain species and encourages the flourishing of incoming others (see White 2013: 462). Eradication may begin, enacted by human and non-human frontier agents, and human agents become instrumental in deliberate campaigns of poisoning and capture, as well as ideological movements through public education, and policy to render species undesirable, threatening, pestly and morally dubious. So too it is often human agents that are responsible for the importation of non-human frontier agents in the form of introduced and potentially invasive (feral) plants and animals. Introduced species can set into action forms of elemental place decay through competition with existing non-human animals and destruction of the nested ecologies into which they are emplaced. Introducing species from elsewhere, can be an attempt to 'make-over' place in the image and styling of other places, a primary strategy in legitimating new ethnic presences and attempting to instate new order over existing.

Accounts of deliberate killing, species decline and ferality chronicle instances across broad time and geographical distances, conveying something of wounding as a shared experience. Deliberate killings are awful events that have occurred in a number of conflict situations, both from individual to group and mass killings. Species decline and suffering, has been recorded as a consequence of all wars, including the Vietnam War (1955–1975), in which the country's jungles and waterways were devastated, and widespread species deaths were induced by carcinogenic poisoning from biological and chemical weapons (Roth-Johnson 2011: 222). During periods of war, the Khmer Rouge also enacted violent environmental policies including extensive timber logging and destruction of ecosystems, which led to widespread deaths and suffering for human and non-human species (Roth-Johnson 2011: 222). More recently, civil war in South Sudan and conflict in Central and West Africa has led to the burgeoning of a bushmeat industry, and killing of wildlife to feed both soldiers and rebels (Dudley et al. 2002: 322; Swamy and Pinedo-Vasquez 2014: 11). Here, the killing and trafficking of non-human animals increases in direct relation to wars, uprisings and displacement (Swamy and Pinedo-Vasquez 2014: 11).

Non-human animal killings in the context of ethnic violence and fighting occurred also during the Bosnian (1992–1995) and Kosovo (1999) wars, as livestock and domestic pets were killed during attacks on villages and cities (Organization for Security and Co-Operation in Europe 1999). Killed as one part of a wider military strategy, these animals were erased in the fashion that human life too was erased. Mass graves containing non-human animal remains, and waterways fouled with their dumped bodies are reported throughout the literature and media accounts of conflict during the mid- and late 1990s (Human Rights Watch 2001; Reitman 1999; ICRC 1999). As a strategy of erasure and inducing fear, the killing (through burning or slaughter) of domestic pets, farm animals and livestock, was horribly effective as a method for wounding place. Furthermore, the Bosnian War, in displacing civilian populations, in the capital of Sarajevo and also in Herzegovina, left countless domestic pets, in particular dogs, without homes. In the years that followed, the problem of stray dogs has accelerated, emerging as a contemporary crisis of dog suffering, animal abuse and the call for official solutions (Drakulic 2014). The notorious 'dogs of Sarajevo' have received international attention and are now the subject of much animal rights debate and animal welfare education.

Other killings precipitated by war and attacks on place concern the demise of those non-human animals held captive in zoos. The perpetrators of violence in such cases have been both the incoming agents of harm and those charged with the safeguarding and survival of such animals. No species is inherently equipped with an ability to withstand violence and trauma, yet for zoo animals, the "inability to anticipate, cope with, and resist and recover from the impact of war is compounded by two factors: the ideology of speciesism that governs nearly all human-animal relations, and the *permanent* nature of zoo animals' institutionalized existence" (Kinder 2013: 53). They are spatially and physically constrained, and their entire lives are subject to almost unmitigated control by humans, rendering them extremely vulnerable to acts of violence in place (Kinder 2013: 53). They are vulnerable to aerial bombardment, looting, and are unlikely to survive if released (Kinder 2013: 53–55). Accounts reveal that during the first Iraq War, animals were bombed, shelled, oiled, gassed, evicted and poisoned as a result of conflict. More than 400 animals were killed or died from starvation, incineration and other injuries in the Kuwait Zoo (Kinder 2013: 54). Countless more were killed by poisoned water, oil explosions and environmental devastation. During the second Iraq War, Baghdad Zoo's population of 250 animals was reduced down to approximately 65, most caused by violent deaths or starvation (Kinder 2013: 71).

Similar narratives of contemporary losses exist for the Kabul, Tripoli and Gaza Zoos, as well as the Sarajevo Zoo. At the Kabul Zoo, it is alleged that during the 1990s, Taliban fighters commandeered the space, "eating the deer, shelling the aquarium and destroying many of the zoo's collection of captive animals" (Kinder 2013: 55). There are historical precedents of mass killings in times of ethnic conflict and war. At the Ueno Zoo, Tokyo, there stands a memorial to the animals killed by Japanese authorities in the days and weeks prior to Chinese air raid strikes in the late 1930s and 1940s (Litten 2009).

Ethnic conflict and the violence of war lead to the deliberate killing of animals, whether as pre-emptive strikes to ensure personal and collective safety on behalf of those under attack, as suggested by events at the Ueno Zoo, or as part of a sweeping strategy to kill elements of place and those non-human animals that feature as part of an order of value and meaning in place. Death and decline come not only from the arrival of harmful human agents, or the impeding threat of conflict between human presences, but are contained in the act of importing new non-human animal species. It is the latter that has figured heavily in ethnographic accounts of place decline across Indigenous homelands in Australia.

Since British colonisation, Australia has suffered a catastrophic loss of native plants and animals, with extinctions numbering well over 100 (more than 50 mammals and birds, 60 plant species and four frogs). An additional 310 species of animals and over 1,180 species of plants are also at risk of disappearing forever (Australian Government Department of the Environment 2004; Ziembicki et al. 2013: 78). These declines and extinctions are not some "now closed and regrettable episode of past colonial history", but rather are an ongoing pattern, with continuing decline (Ziembicki et al. 2013: 78). Colonialism, in Australia and also across the Americas and New Zealand, supported the slaughter of native animals and the importation of introduced species. The reason for this practice, according to White (2013: 463), is traceable to some of the following:

> native animals could get in the way of pastoral farming, the strangeness of
> the animals made them easier to demonise; the countryside, to European

eyes, could be barren, monotonous, unreadable and truly wild; lacking the familiar sights and sounds of Britain, homesickness was ever-present.

Belcourt (2015: 1) contends that in a settler colonial context, such as North America, "animal domestication, speciesism, and other modern human-animal interactions ... are possible because of and through the erasure of Indigenous bodies and the emptying of Indigenous lands for settler-colonial expansion". Settler colonialism and white supremacy as "political machinations" require the "simultaneous exploitation and/or erasure of animal and Indigenous bodies". It is such that animality becomes "a politics of space", through which "animal bodies are made intelligible in the settler imagination on stolen, colonized, and re-settled Indigenous lands". The politics of space that Belcourt (2015: 3) articulates marries into the practice of colonialism's violence in place; as violence is a political act; as an exercise in or expression of power. How the "settler moves to knowing and/or constructing animal bodies and/or subjectivities (re)locates animals within particular geographic and architectural spaces". That this may call upon acts of decimation, and erasure is part of the aggressive politicking of relational encounters that involve the taking of place and killing or diminishing of existing life.

A wide range of species have been introduced into Australia, beginning with pigs and cattle followed by horses, donkeys, buffalo and camel. Some birds, such as English songbirds, were imported for their musical reminders of home, while rabbits and foxes were introduced so as to be hunted by settlers (McLeod 2004, cited in White 2013: 461). Countless introduced species have escalated to 'invasive' status, for the threat they pose to 'native' biodiversity. Distinguished as invasive, feral or pest, this language of distinction echoes Belcourt's (2015: 3) point as to the knowing and constructing of animal bodies within particular geographies and architectural spaces. This language 'of what belongs' (see Trigger 2008) is striking in Australia, where discourse on ferality slips from non-human animals to human presences very easily, revealing "links in Australia between ideas of 'nativeness' in society and nature" (Trigger 2008: 628). Trigger also explains that, in Australia:

> The areas that have been penetrated by weeds and feral animals are less 'healthy'. From the perspective of cultural analysis, such figurative depictions of the ecological impacts of European settlement immediately prompt parallels with maps showing the history of impacts on Aboriginal people. On the 'landscape health' maps, a line divides eastern and southern Australia (an 'intensive use zone' that has depleted native environments) from northern and most of Western Australia (which still has more than 90 per cent native flora). And this line largely mirrors a designation of historically more intense socio-cultural impacts upon Aboriginal groups in the south and east as compared with those whose

languages and cultural traditions have remained less affected by the broad process of European settlement (Broome 1982: 37; Rowley 1972: xiv–xv).
(Trigger 2008: 629)

In sum, as species numbers and ecological order decline, and as elemental decay sets in, then entire lifeways (human and non-human) may also fall victim to agents of harm and the greater effects of cultural wounding and moral disengagement. Hence, it is such that elemental decay, in the form of deliberate killing, species decline and introduced ferality are often triumphant in their delivery of place harm and deeply felt cultural wounding. It is, as proposed in the opening paragraph of this chapter, the logic of death from below and all around.

Concluding remarks

This chapter and the two preceding have brought forth the realisation that there are a number of ways in which place might be harmed and subsequently wounded. So too is there a distinction in the human as an agent of harm. In coming to know place harm, as brought about by human agents, the first distinction comes from seeing its deliberacy. The intent to harm requires particular epistemological, ontological and axiological dispositions. First there is the denial of kinship between people and place (as constituted by a nested ecology of multifarious agents and co-presences), followed by the intent to place particular forms of human life above all else. These two combined ensure that place becomes a negative externality, upon which violence can be acted. Once this ideological habit has been perfected, then axiological retreat and disregard become part of a wider moral disengagement that allows awful events to occur in place, and assaults to take place that target the nested ecologies of the personal, social, environmental, cosmic and spiritual realms of life and the living. These ecologies, however, do not only matter when human life is found at their centre, rather they must be understood as incorporating all living 'things' as components of the relational world. How well these are articulated depends on the cultural context and epistemic space in which one operates or to which one might be exposed and receptive.

This book in its commitment to Indigenous epistemologies of place and decolonising methodologies, positions the human as but one element in and of place, located among a myriad of interacting co-presences. Interacting co-presences may include the human, but do not prioritise this single element of place, nor position it above other presences. Place is what brings all of these agents into a nested practice of relating. Of this nesting, I introduce Lynch (1972: 65), whose words contextualise further the tension that comes from centralising human life and human suffering; a tendency that leads to an oversight of relationality and therefore consciousness of other suffering that comes with violence and wounding in place. Change – as brought about by human violence and disassociation with the needs and order of other worldly presences – denies the human its place within a nested ecology. It outrageously seeks to distinguish the human as capable of invading all realms of life. Yet the place world disrupts this fantasy time and again, through its own agency and sentiency. That its presence looms large, even in the most ghastly of wounding acts is now clearer to me than ever before. Any attempt to naturalise what is done to the place world as a result of human conflict, or assumption that place somehow withstands the brutality of war, ethnic conflict, mining, toxicity and elemental decay, is withered, for as Lynch reflects:

> We have two kinds of evidence for the passage of time. One is rhythmic repetition – the heartbeat, breathing, sleeping and waking, hunger, the cycles of sun and moon, the seasons, waves, tides, clocks. The other is progressive and irreversible change – growth, decay, not recurrence but alteration. Men have made magical attempts to see the second phenomenon as a cosmic variant of the first, to pretend that change is also cyclical, to imagine that progressive time is a series of eternal, contrasting repetitions, each arising from the other. That magic warms the spirit with the sense that decline and dissolution are only appearances that resurrection will follow. But the things we love do not in fact come back to us. Whatever our hopes, we know things change.
>
> (Lynch 1972: 65)

This chapter, along with Chapters 3 and 4, has sought to describe a spectrum of experience across which place is harmed and made subject to the greater effects of cultural wounding. As nuanced as the human response might be to cultural wounding and violence in conflict situations, place also is nuanced in its response, as both decline and survival. There are many more questions on place survival and place healing that arouse my curiosity and to these I feel compelled to now turn.

Notes

1 Ecocide occurs as the result of large-scale destruction caused by aggressive land use or land-clearing processes, pollution and dangerous industrial activity (Higgins 2010). In 2014, a submission was made to the International Criminal Court at the United Nations to include 'The Crime of Ecocide' under Article 5 of the Rome Statute (Harvey 2012; Higgins 2010). This would add ecocide to the list of Crime[s] Against Peace, including the Crime of Genocide, Crimes Against Humanity, War Crimes and Crimes of Aggression (Harvey 2012; Higgins 2010). The inclusion of ecocide law as international law prohibits mass damage and destruction of the earth and creates a legal duty of care for all inhabitants that have been or are at risk of being significantly harmed due to ecocide. Currently there is no overriding mandatory duty of care (fiduciary duty) to prohibit, prevent significant hazards or harm, or to pre-empt by assisting to those who are facing ecocide (Higgins 2010).

2 A nuclear claims tribunal was established in 1988 to assess personal injury claims throughout the archipelago, as cancers and thyroid complaints have increased (DeGroot 2004: 330). Assessments of tumours, leukaemia, cancer, birth defects and

overall suffering have been heard. The Bikinians have also launched a separate claim against the US government, which agreed to pay damages of nearly $565 million as a final settlement of all claims (DeGroot 2004: 330). Since the 1990s, the atoll has also opened for business, with diving, sport fishing and tourism striving to redefine the region (De Groot 2004: 331). And while it is now considered 'radiologically safe for visitation' the International Atomic Energy Agency "does warn that 'eating locally grown produce, such as fruit could add significant radioactivity to the body'" (DeGroot 2004: 331). While apparently posing no threat to tourism, long-term habitation is untenable.

References

Abel, T. and Stepp, J.R. 2003. A New Ecosystems Ecology for Anthropology. *Conservation Ecology* Vol. 7(3): 12. Available at www.ecologyandsociety.org/vol7/iss3/art12.

Abram, D. 1996. *The Spell of the Senuous*. New York: Random House.

Abram, D. 2011. *Becoming Animal*. London: Vintage Books, Random House.

Australian Government Department of the Environment. 2004. *Threatened Species and ecological Communities in Australia.* Available at www.environment.gov.au/biodiversity/threatened/publications/threatened-species-and-ecological-communities-australia. Accessed 17 November 2015.

Barker, H. 2004. *Bravo for the Marshallese: Regaining Control in a Post-Nuclear, Post Colonial World*. Belmont, California: Wadsworth.

Bartholet, J. 1997. Bikini Atoll's Blue Lagoon. *Newsweek* (International Edition), 12 December.

Beck, U. 1999. *World Risk Society*. Malden, Massachusetts: Polity Press.

Bekoff, M. 2007. *The Emotional Lives of Animals*. California: New World Library.

Bekoff, M. 2010. *The Animal Manifesto: Six Reasons for Expanding Our Compassion Footprint*. California: New World Library.

Bekoff, M. 2014. *Rewilding Our Hearts: Building Pathways of Compassion and Coexistence*. California: New World Library.

Belcourt, B. 2015. Animal Bodies, Colonial Subjects: (Re)Locating Animality in Decolonial Thought. *Societies* Vol. 5: 1–11.

Blades, D. and Siracusa, J. 2014. *A History of US Nuclear Testing and its Influence on Nuclear Thought, 1945–1963*. New York: Rowman and Littlefield.

Bonner, K. 1997. *Final Voyages*. New York: Turner Publishing.

Bradley, J. 2008. When a Stone Tool is a Dingo: Country and Relatedness in Australian Aboriginal Notions of Landscape, in *Handbook of Landscape Archaeology*. Edited by B. David and J. Thomas, pp. 633–637. Walnut Creek, California: Left Coast Press.

Bridges, K., and McClatchey, W. 2009. Living on the Margin: Ethnoecological Insights from Marshall Islanders at Rongelap Atoll. *Global Environmental Change* Vol. 19: 140–146.

Broome, R. 1982. *Aboriginal Australians*. North Sydney: Allen & Unwin.

Brown, S. 2013. Archaeology of Brutal Encounter: Heritage and Bomb Testing on Bikini Atoll, Republic of the Marshall Islands. *Archaeology in Oceania* Vol. 48: 26–39.

Creative Spirits. nd. *Maralinga: How British Nuclear Tests Changed History Forever*, Available at www.creativespirits.info/aboriginalculture/history/maralinga-how-british-nuclear-tests-changed-history-forever#ixzz3oUTqqWpH. Accessed 14 October 2015.

CTBTO. nd. *Nuclear Testing, World Overview*. Available at www.ctbto.org/nuclear-testing/history-of-nuclear-testing/world-overview. Accessed 16 September 2015.

Davis, J. 2005a. Representing Place: 'Deserted Isles' and the Reproduction of Bikini Atoll. *Annals of the Association of American Geographers* Vol.95(3):607–625.

Davis, J. 2005b. "Is It Really Safe? That's What We Want to Know": Science, Stories, and Dangerous Places. *The Professional Geographer* Vol. 57(2): 213–221.

DeGroot, G. 2004. *The Bomb: A Life*. London: Random House.

Drakulic, S. 2014. The Dogs of Sarajevo. *Eurozine*, 21 January. Available at www.eurozine.com/articles/2014-01-21-drakulic-en.html. Accessed 20 October 2015.

Dudley, J., Ginsberg, J., Plumptre, A., Hart, J. and Campos, L. 2002. Effects of War and Civil Strife on Wildlife and Wildlife Habitats. *Conservation Biology* Vol. 16(2): 319–329.

Erikson. K. 1976. *Everything in its Path: Destruction of Community in the Buffalo Creek Flood*. New York, London: Simon and Schuster Paperbacks.

Erikson, K. 1994. *A New Species of Trouble: Explorations in Disaster, Trauma and Community*. New York: W.W. Norton and Co.

Erikson, K. 2011. The Day the World Turned Red: A Report on the People of Utrik. *Yale Review* Vol. 99(1): 27–47.

Fullilove, M. 2004. *Root Shock: How Tearing Up City Neighborhoods Hurts America and What We Can Do About It*. New York: Ballantine Books.

Goodall, H. 1994. Colonialism and Catastrophe: Contested Memories of Nuclear Testing and Measles Epidemics at Ernabella, in *Memory and History in Twentieth-Century Australia*. Edited by K.D. Smith and P. Hamilton, pp. 55–76. Melbourne: Oxford University Press.

Grosfoguel, R. 2007. The Epistemic Decolonial Turn: Beyond Political-Economy Paradigms. *Cultural Studies* Vol. 21(2–3): 211–223.

Guyer, R. 2001. Radioactivity and Rights: Clashes at Bikini Atoll. *The American Journal of Public Health* Vol. 91(9): 1371–1376.

Harvey, G. 2005. *Animism: Respecting the Living World*. London: Hurst, and New York: Columbia University Press.

Harvey, G. 2006. Animals, Animists, and Academics. *Zygon* Vol. 41(1): 9–20.

Harvey, R. 2012. 'Ecocide' – Will This be the Fifth International Crime Against Peace? *Socialist Lawyer* Vol. 61: 11.

Hediger, R. 2013. *Animals and War: Studies of North America and Europe*. Leiden, Netherlands: Brill.

Heider, F. 2005. Violence and Ecology. *Peace and Conflict: Journal of Peace Psychology, Special Issue: Military Ethics and Peace Psychology: A Dialogue* Vol. 11(1): 9–15.

Higgins, P. 2010–2015. Eradicating Ecocide: What is Ecocide? Available at http://eradicatingecocide.com/the-law/what-is-ecocide. Accessed 12 October 2015.

Hinton, A.L. (ed.). 2002. *Annihilating Difference: The Anthropology of Genocide*. Berkeley, California: University of California Press.

Human Rights Watch. 2001. *Under Orders: War Crimes in Kosovo*, Chapter 4, March–June 1999: An Overview. Available at www.hrw.org/reports/2001/kosovo/undword-03.htm. Accessed 19 October 2015.

Hyndman, D. 1988. Ok Tedi: New Guinea's Disaster Mine. *Ecologist* Vol. 18: 24–29.

ICRC. 1999. Balkan Crisis: Cleaning Wells in Kosovo. International Committee of the Red Cross, News 99/37, 8 September 1999. Available at www.icrc.org/eng/resources/documents/misc/57jpzc.htm. Accessed 19 October 2015.

Jasanoff, S. 1999. The Songlines of Risk. *Environmental Values* Vol. 8(2): 135–152.

Jenkins, M. 2008. Who Murdered the Virunga Gorillas? *National Geographic* Vol. 214(1): 34–65.

Johnstone, B. and Barker, H. 2008. *The Rongelap Report: Consequential Damages of Nuclear War*. Walnut Creek, California: Left Coast Press.

Jose, C., Wall, K. and Hinzel, J.H. 2015. This Dome in the Pacific Houses Tonnes of Radioactive Waste and it's Leaking. *The Guardian*, 3 July 2015. Available at www. theguardian.com/world/2015/jul/03/runit-dome-pacific-radioactive-waste. Accessed 14 September 2015.

Karrabing. 2015. *Keeping Country Live, Media Projects*. Available at www.karrabing. com. Accessed 17 November 2015.

Kearney, A. 2014. *Cultural Wounding, Healing and Emerging Ethnicities: What Happens when the Wounded Survive?* New York: Palgrave Macmillan.

Kearney, A. and Bradley, J. 2015. When a Long Way in a Bark Canoe Becomes a Quick Trip in a Boat: Changing Relationships to Sea Country and Yanyuwa Watercraft Technology. *Quaternary International* Vol. 385: 166–176.

Kinder, J. 2013. Zoo Animals and Modern War: Captive Casualties, Patriotic Citizens and Good Soldiers, in *Animals and War: Studies of Europe and North America*. Edited by R. Hediger, pp. 45–76. Leiden, Netherlands: Brill.

Kirsch, S. 2007. Indigenous Movements and the Risks of Counter Globalization: Tracking the Campaign Against Papua New Guinea's Ok Tedi Mine. *American Ethnologist* Vol. 34(2): 303–321.

Leschine, T. 2014. Risk and Vulnerability at Contaminated Sites in the Pacific and Australian Proving Grounds from a 'Long-term Stewardship' Perspective: What Have We Learned? in *The Oceans in the Nuclear Age: Legacies and Risks*. Edited by D. Caron and H. Scheiber, pp. 39–48. Boston: Leiden.

Levi, J.M. 1993. *Pillars of the Sky: The Genealogy of Ethnic Identity among the Raramuri-Simaroni (Tarahumara Gentiles) of Northwest Mexico*. Dissertation Harvard University, Cambridge, Massachusetts, USA.

Litten, F.S. 2009. Starving the Elephants: The Slaughter of Animals in Wartime Tokyo's Ueno Zoo. *The Asia-Pacific Journal* Vol. 7(38), Number 3.

Low, N. and Gleeson, B. 1998. Situating Justice in the Environment: The Case of BHP at the Ok Tedi Copper Mind. *Antipode* Vol. 30(3): 201–226.

Lynch, K. 1972. *What Time is This Place?* Massachusetts: Massachusetts Institute of Technology.

Maragos, J. 2011. Bikini Atoll, Marshall Islands, in *Encyclopedia of Modern Coral Reefs*. Edited by D. Hopley, pp. 123–136. Netherlands: Springer.

McLeod, R. 2004. *Counting the Cost: Impact of Invasive Animals in Australia 2004*. Cooperative Research Centre for Pest Animal Control. Canberra. Available at www. pestsmart.org.au/wp-content/uploads/2010/03/CountingTheCost.pdf. Accessed 19 October 2015.

Muir, C., Rose, D.B. and Sullivan, P. 2010. From the Other Side of the Knowledge Frontier: Indigenous Knowledge, Social-ecological Relationships and New Perspectives. *The Rangeland Journal* Vol. 32(3): 259–265.

Navy History and Heritage Command. 2002. Operation Crossroads: Fact Sheet. Washington, DC. Available at www.history.navy.mil/research/library/online-reading-room/title-list-alphabetically/o/operation-crossroads/fact-sheet.html. Accessed 8 October 2015.

Niedenthal J. 1997. A History of the People of Bikini Following Nuclear Weapons Testing in the Marshall Islands: With Recollections and Views of Elders of Bikini Atoll. *Health Physics* Vol. 73(1): 28–36.

Nixon, R. 2011. *Slow Violence and the Environmentalism of the Poor.* Cambridge: Harvard University Press.

Organization for Security and Co-Operation in Europe. 1999. *Kosovo/Kosova: As See, As Told.* An analysis of the human right findings of the OSCE KosovoVerificiatio Mission, October 1998 to June 1999. Warsaw, Poland: Organization for Security and Co-operation in Europe, Office for Democratic Institutions and Human Rights. Available at www.osce.org/odihr/17772?download=true. Accessed 19 October 2015.

OTML. 2005. *CMCA Environmental Predictions Update.* Tabubil, Papua New Guinea: Ok Tedi Mining, Ltd.

Peeples, J. 2011. Toxic Sublime: Imaging Contaminated Landscapes. *Environmental Communication* Vol. 5(4): 373–392.

Plumwood, V. 2002. *Environmental Culture: The Ecological Crises of Reason.* London: Routledge.

Plumwood, V. 2012. *The Eye of the Crocodile.* Edited by L. Shannon. Canberra: Australian National University E Press.

Povinelli, E. 1993. 'Might Be Something': The Language of Indeterminacy in Australian Aboriginal Land. *Man* Vol. 28(4): 679–704.

Povinelli, E. 2014. Elizabeth Povinelli Website and Blog. Available at http://elizabethpovinelli.com/blog. Accessed 12 October 2015.

Reitman, V. 1999. Kosovo Wells Emerging as Mass Graves. *Los Angeles Times,* 10 August. Available at http://articles.latimes.com/1999/aug/10/news/mn-64278. Accessed 19 October 2015.

Ricard, M. 2013. *Altruism: The Power of Compassion to Change Yourself and the World.* London: Atlantic Books.

Ricard, M. 2015. *In Defense of Animals.* Available at www.matthieuricard.org/en/blog/posts/in-defense-of-animals. Accessed 20 October 2015.

Rose, D.B. 1992. *Dingo Makes Us Human: Life and Land in an Australian Aboriginal Culture.* Cambridge: Cambridge University Press.

Rose, D.B. 1996. *Nourishing Terrains: Australian Aboriginal Views of Landscape and Wilderness.* Canberra: Commonwealth of Australia.

Rose, D.B. 2000. *Dingo Makes Us Human: Life and Land in Australian Aboriginal Culture* (2nd edition). Cambridge: Cambridge University Press.

Rose, D.B. 2004. *Reports from a Wild Country: Ethics for Decolonisation.* Sydney, New South Wales: UNSW Press.

Rose, D.B. 2008a. On History, Trees, and Ethical Proximity. *Postcolonial Studies* Vol. 11(2): 157–167.

Rose, D.B. 2008b. Fitting into Country: Ecology and Economics in Indigenous Australia. *Capitalism, Nature, Socialism* Vol. 19(3): 117–121.

Rose, D.B. 2008c. Dreaming Ecology: Beyond the Between. *Religion and Literature* Vol. 40(1): 109–122.

Rose, D.B. 2011. *Wild Dog Dreaming: Love and Extinction.* Charlottesville and London: University of Virginia Press.

Rose, D.B. 2013. Death and Grief in a World of Kin, in *The Handbook of Contemporary Animism Edition One.* Edited by G. Harvey, pp. 137–147. Durham: Acumen.

Rose, D.B. 2014, Arts of Flow: Poetics of 'Fit' in Aboriginal Australia. *Dialectical Anthropology* Vol. 38(4): 431–445.

Roth-Johnson, D. 2011. Human Geography, in *Green Culture.* Edited by K. Wehr and P. Robbins, pp. 219–226. Los Angeles and London: Sage.

Rowley, C. 1972. *Outcasts in White Australia.* Ringwood, Victoria: Penguin.

Salmón, E. 2000. Kincentric Ecology: Indigenous Perceptions of the Human–Nature Relationship. *Ecological Applications* Vol. 10(5): 1327–1332.

Shkilnyk, A. 1985. *A Poison Stronger than Love: The Destruction of an Ojibwa Community.* New Haven: Yale University Press.

Silko, L.M. 1996. *Yellow Women and a Beauty of the Spirit.* New York: Simon and Shuster.

Stanner, W.E.H. 1965. Religion, Totemism, and Symbolism, in *Aboriginal Man in Australia: Essays in Honour of Emeritus Professor A.P. Elkin.* Edited by R. Berndt and C. Berndt, pp. 207–237. Sydney: Angus & Robertson.

Swamy, V. and Pinedo-Vasquez, M. 2014. *Bushmeat Harvests in Tropical Forests: Knowledge Base, Gaps and Research Priorities.* Occasional paper 114, Center for International Forestry Research. Bogor, Indonesia: CIFR.

Tabucanon, G.M. 2014. Protection for Resettled Island Populations: The Bikini Resettlment and its Implications for Environmental and Climate Change Migration. *Journal of International Humanitarian Legal Studies* Vol. 5: 7–41.

Takada, J. 2000. *Radiological Investigations in Rongelap Island 1999.* First report prepared for the People of Rongelap as a humanitarian support, October, 2000. International Radiation Information Center, Research Institute for Radiation Biology and Medicine, Hiroshima University. Available at www.moruroa.org/medias/pdf/Rapport%20Rongelap%201999.pdf. Accessed 3 July 2015.

Tamisari, F. 2002. Names and Naming: Speaking Forms into Place, in *The Land is A Map: Placenames of Indigenous Origin in Australia.* Edited by L. Hercus, F. Hodges and J. Simpson, pp. 87–102. Canberra: Australian National University E Press.

Titus, C.A. 2001. *Bombs in the Backyard Atomic Testing and American Politics.* Reno: University of Nevada.

Tobin, J. 2002. *Stories from the Marshall Islands: Bwebwenato Jān Aelōñ Kein.* Honolulu: University of Hawaii Press.

Totten, S., Parsons, W. and Hitchcock, R. 2002. Confronting Genocide and Ethnocide of Indigenous Peoples: An Interdisciplinary Approach to Definition, Intervention, Prevention and Advocacy, in *Annihilating Difference: The Anthropology of Genocide.* Edited by A.L. Hinton, pp. 54–94. Berkeley, California: University of California Press.

Townsend, W. 1988. Giving Away the River: Environmental Issues in the Construction of the Ok Tedi Mine, 1981–1984, in *Potential Impacts of Mining on the Fly River.* Edited by J. Pernetta, pp. 107–119. United Nations Environment Program: UNEP Regional Seas Reports and Studies, 99, and SPREP Topic Review, 33. Geneva: UNEP.

Trigger, D. 2008. Indigeneity, Ferality and What 'Belongs' in the Australian Bush: Aboriginal Responses to 'Introduced' Animals and Plants in a Settler-Descendant Society. *Journal of the Royal Anthropological Institute* Vol. 14: 628–646.

Trumbull, R. 1982. An Island People Still Exhiled by Nuclear Age: Natives of Bikini Want Only to Return to their Pacific Paradise. *US News and World Report*, 18 October: 48–50.

Villagra, A. 2011. Cannibalism, Consumption, and Kinship in Animal Studies, in *Making Animal Meaning.* Edited by L. Kalof and G. Montgomery, pp. 45–58. East Lansing, Michigan: Michigan State University Press.

Viveiros de Castro, E. 1996. Images of Nature and Society in Amazonian Ethnology. *Annual Review of Anthropology* Vol. 25: 179–200.

Viveiros de Castro, E. 1998. Cosmological Deixis and Amerindian Perspectivism. *The Journal of the Royal Anthropological Institute* Vol. 4(3): 469–488.

White, S. 2013. British Colonialism, Australian Nationalism and the Law: Hierarchies of Wild Animal Protection. *Monash University Law Review* Vol. 39(2): 452–472.

Wimberley, E. 2009. *Nested Ecology: The Place of Humans in the Ecological Hierarchy.* Baltimore: John Hopkins University Press.

Winkler, A. 1999. *Life Under a Cloud: American Anxiety about the Atom.* Oxford: Oxford University Press.

Ziembicki, M., Woinarski, J. and Mackey, B. 2013. Evaluating the Status of Species Using Indigenous Knowledge: Novel Evidence for Major Native Mammal Declines in Northern Australia. *Biological Conservation* Vol. 157: 78–92.

Part IV

Reinstating kinship and healing place

The final section of this book, Part IV, is directed towards consideration of the important questions, can places heal? How might they heal? Seeking pathways to an axiological return, in which kinship between people and place is found and inspired, is the final objective here. This requires a reflective discussion on the principle of kincentric ecology, as a heuristic and practical device aimed at redressing the place harms that occur worldwide. Most importantly, the challenge of how these principles can be lived and entered into the normative practice of everyday place interactions among cultural and ethnic groups, as inclusive of Indigenous and non-Indigenous identities, is addressed.

A discussion of healing is timely and necessary at this point, as a commitment to finding pathways out from wounding and away from compulsions to harm. This final discussion strives to better understand the role that human life might play in mitigating harm and piecing together the broken or disordered elements of place that ultimately cause broken and disordered cultural and social lives. The practical aspects of living with kincentricity and within a nested ecology are considered more closely. It is by reinstating kinship that ontological shifts might occur. This is not a naïve assertion unbecoming of neoliberal life, nor is it cast as an esoteric solution to a practical 'problem'. Rather this is part of a wider movement already well underway for millennia, in the form of Indigenous epistemologies, ontologies and axiologies around the world. So too it shows itself in the workings of emotional geography, which has encouraged epistemological openness in how non-Indigenous thinkers begin to question the ways in which we see the world and understand place.

6 Kincentric ecology and seeking an axiological return

This book commits to recognition of deep connections and profound relationality in the life world as something that enriches the scope of our cultural geographies. When aligned with Indigenous epistemologies of place, this approach operates on the principle that we (humans) are nested within a relational sphere that is populated by a vast number of co-presences, each with the capacity for agency and sentiency. These co-presences, when understood as kin, enter into our axiological frame and emotional geography in ways that cannot be ignored. Neither agency nor sentiency are positioned as uniquely human qualities, instead they are faculties possessed by place itself, and place elements. Although few within social and cultural geography would cast place as a blank canvas painted to life solely by human imagination, or action, the push to more fully understand place as a spatial and spectral (essence-filled) element of life, populated by a vast number of multifarious agents that respond to human life remains a committed one.

One approach to this involves exploring more fully the impact of locating place within kincentric ecology. Kincentricity is a key aspect of Indigenous worldviews, yet here it is epistemologically and ontologically expanded and applied to contexts that also include non-Indigenous place relations and experiences of violence. By adopting a kincentric ecological approach, this book has sought possibilities for a more far-reaching axiological return to place. The principle of nesting amid kin and with important relational others (co-presences in place) is returned to as part of a deep and moral ecology of place relations. Delineating a kincentric ecology on terms that can be grasped more fully from within the disciplinary praxis of social and cultural geography and also within an anthropology of place is demonstrative of decolonising methodologies that disrupt and decentre Western epistemologies as necessarily enough to contend with the impact of human conflict, violence and cultural wounding on the place world.

As an anthropologist, it has become clear to me that there are diverse epistemic spaces in which place is configured, yet the West and its epistemologies (as part of a colonial power matrix) have tended to dominate accounts of how people interact with place, generating a normative view of how we configure place, interact with it and potentially regard or disregard its wounding. With growing dissatisfaction of what these normative views of place entail, it is to the epistemic spaces of Indigenous kincentric ecologies that this work is ultimately drawn. And it is from here that this work finds its most substantive insight in seeking an axiological return to place value.

Many Indigenous epistemologies rely on a kincentric way of being-in-relation. According to Salmón (2000) and Bradley (with Yanyuwa Families 2016) this way of being is encoded in language and represents a worldview, a way of thinking and overall cognition. Not only affecting the way people think, a kincentric way of being brings ontological and axiological specificity, namely it affects the kinds of people we are and aspire to be. Kincentricty is interdependency, whereby the survival of people is intertwined with that of place and place elements and vice versa. Place is dependent upon people to ensure its survival both in material and ecological terms, as well as social and spiritual terms. While there are many cultural instances in which claims of this kind may be disputed, there are a great number of cultural contexts in which relational co-presences in place are indivisible from their human kin, and contribute vital elements to the greater whole that is life. The key to this final chapter is finding a discourse to expand the scope of kincentric ecology, so that it might be proposed as a necessary return to the 'whole way of life' that shapes people and place relations, and how the greater effects of cultural wounding on place are understood and redressed.

Bhattacharyya and Slocombe (2014) identify key elements to a kincentric ecological approach, including, a "shift in emphasis from human rights to human responsibilities to wildlife", "a focus on system dynamics rather than just specific species", "the acknowledgement of uncertainty and rapid social-ecological change" and "an emphasis on the value of locally relevant, empirical knowledge to effective management" (Bhattacharyya and Slocombe 2014). These inspire the first step in this critical discussion, which locates the greater effects of cultural wounding, and the impact of human violence on place, within a kincentric ecological approach. Bhattacharyya and Slocombe (2014) inspire the critical points, which guide an axiological return, yet these have been modified to fit the specificity of a place discussion. Here, they include, 'human responsibility over human rights', 'relational dynamics as affected by uncertainty and rapid change' and 'local empiricism and place order as mechanisms for healing'. Configuring people and place relations through this chain of reflection and understanding is what highlights the concatenation of circumstances that might ultimately lead to axiological return.

Human responsibility over human rights

In the opening pages of this book the question was asked, what is it that we, as humans, owe the place world? Where disorder and the unsettling conditions of place harm take hold, precipitated by the effects of cultural wounding and its associated axiological retreat, human life is drawn into a relational encounter where response and non-response might appear to be two options.

Responsive reflexivity is juxtaposed by unresponsive reflexivity, yet exists as one part on a continuum of engagement. Here reflexivity is treated as that state of being in which the self is encountered as a communal and individual actor, consistently problematised in relation to something or someone else. The self is constructed as a participant, operating within certain structures and dispositions that relate to and shape the way we know, act, care, perceive and configure the world. Reflexivity is the process of coming to know and speak or write of these structures and dispositions and how they affect the being that we are (Finlay 2003; Gough 2003).

Responsive reflexivity involves fully engaging reflexive self-awareness, and in doing so acknowledging reflexivity beyond the self, in which case the condition of knowing the self increases the likelihood of seeing and knowing the existences of others, human or otherwise. In the case of violence and harm-doing, this reflexive awareness is extended to imagine and more fully witness the experiences of inflicted harm, which are lived through by all other co-presences. Unresponsive reflexivity stops at the self, and denies an expanding reflexive awareness to include other beings and dispositions of consequence and importance. In sum, it manifests as unwillingness to imagine the lived experience of violence, a lack of commitment to witnessing such events and lingering encounters, and sadly, a denial of violence and its harmful effects altogether.

In the first instance, namely responsive reflexivity, responsiveness is receptiveness to the acknowledgement of harm and sensitivity to its causation. This might be expressed as an empathic response or drive to remediate and mitigate against further harm, through action or ideological position. This is expressed by Indigenous groups through the fight for land rights and their resistance, along with non-Indigenous supporters, to violent extractive industries across much-valued homelands. Responsive reflexivity is performed as the cultural trauma that drives the search for mass graves in ethnically cleansed villages, and is what holds to ground the desperate search for answers to questions of toxicity in former nuclear testing sites. As a response, it heavily involves and implicates the human in the causation of harm, while drawing on human witnesses (or kin) to acknowledge and therefore through response alleviate the suffering of place. This response instates human responsibility and returns human life to a nested ecology in which the effect of harm in one realm threatens harms in another or can be intervened in by the actions of another.

Unresponsive reflexivity is characterised by something different, namely passivity, even in the face of exposure to or knowledge of place harm. As an act of choice, unresponsive reflexivity represents a distant form of witnessing, whereby a person may avail themselves to the communicative event that is place harm to a lesser or greater degree, or may opt out from a dialogic encounter as one that prompts reflection. The witness may be unmoved to care or act in response, instead availing themselves only to the position of 'knowing about' place harm. Such a position may cause an individual or group of people to locate place harm as secondary to the interests of people or inevitable and part of a 'modern world system'. It may be cast as an unfortunate casualty of war, or a necessary burden to be carried by distant cultural others. Human rights overshadow human responsibility in this case, albeit as narrowly conceived in relation to not all humans, but those for whom place harm brings some form of benefit. The human right is configured as the right to command the world around (such as through mining and resource extraction, toxicity, annihilation and designification), as a depoliticised event with naturalised impact. Value nihilism can take hold as a failure to care, or decisive care is mediated only by the effect to which self-interest is under threat, but ultimately the consequential harms are diffused enough so as to restrict the likelihood of existential shock or mute the effects that might warrant human attention.

In spite of their distinctive conditions, responsive reflexivity and unresponsive reflexivity both involve consciousness of actions, events and even change, as a form of witnessing. Yet, these actions and harms are processed through very different axiologies and may lead to very different ontologies in people and place relations. Responsive and unresponsive reflexivity equally invoke moral positions, whereby the decision to care exists at one point of the spectrum while failure to care occupies another (see Scheper-Hughes 1995: 419, for a parallel discussion on 'non-involvement' as an "ethical and moral position"). Whichever the response, witnessing as a nuanced act, compresses or contracts the human into a place relation, as a closer, more intimate encounter, or as a distant form of spectatorship. Witnessing is at once conceived of as 'seeing' something, sitting with an event and its aftermath, having knowledge of place harm or having faith in the claim that something of consequence has occurred.

It is the distinction between 'having knowledge of' and 'having faith in the claim of consequence' that is pertinent to this discussion of human responsibility. To 'have knowledge of place harm', is akin to spectatorship; where witnessing involves the knowledge or even perhaps observation of harm, yet an enduring separation between the effects of that harm and the continuation of certain forms of human life. Consequential harms may be overlooked, justified, diminished or even denied by spectators in this mode of unresponsive reflexivity. Witnessing of the intimate kind and 'having faith' in the occurrence of harm brings an altogether different response. This may be a vivid awareness of circumstances or it may impel revulsion, whereby the human fully participates in the traumatic effects of place harm.

Intimate forms of witnessing embroil human life in the consequences and effects of place harm. This is seen in the case of physical sickness, social decline, root shock and the melancholia that takes hold among human groups as the places they rely upon or participate in are disordered. Ontologically, people come to feel and embody the harm, while seeking out epistemological frameworks for describing it. So too they are axiologically drawn to care and thus express dissent, anger, sadness or longing at what has occurred or continues to occur. Intimate forms of witnessing have at their core a sense of responsibility. This is distinct from a sense of human rights, which locates human well-being as primary. Responsibility throws open the limits of obligation, care, culpability and investment in something greater than human life.

Conversely, human rights address urgent and important moral concerns that are practically or potentially relevant to humans (Yechiel 2012: 149). A dominant discourse is that, whenever possible, respect for human rights must be enforced, even at the expense of other interests and values, and human rights violations are considered intolerable regardless of the social context or excuses given (Yechiel 2012: 149). While this discussion, which promotes shifting the emphasis from human rights to human responsibilities in the context of cultural wounding and associated place harms, is not intended to dilute the importance of recognising and committing to the alleviation of human suffering, it contends that the primacy of human rights as a largely Western discourse is problematic. It is the emphasis on human experience and rights as both 'universally understood' and of overwhelming importance, which elevates human encounters with violence and trauma above all else.

The configuring of rights as distinctively human (with the exception of 'animal rights' discourse, see Regan 2004; Singer 2011) supports a separation between human and 'other' life. The 'all else' logic, which this 'other' life distinction imbibes, falls outside of the axiological range of human rights by its very definition. The 'other' becomes the place world and place elements, including non-human animals, ancestors, weather, geological features and all range of living forms. Somehow this continues, despite knowledge that human action has far-reaching consequences. Where human action is prefaced by the intent to harm and where cultural wounding and violence are enacted, the greater effects may lodge deep within place. As this book reveals, diagnosing place harm, as prompted by acts of violence designed to culturally wound, can be both straightforward (as seen with aerial bombardment, place annihilation and designification) and complicated (as with toxicity and the invisible and sometimes creeping effect of elemental decay in place). The emphasis on human rights maintains the illusion of human primacy, and when facilitated by unresponsive reflexivity, diminishes awareness of and therefore responsibility taken for the effects of human violence, and its capacity to induce deep harms in place.

One way of shrinking the gap, within which axiological crises and failures to care find their hold, is to balance the concern for human rights with that of human responsibility or radically replace the former with the latter as a new framework for configuring rights, accountability, and action. According to Yechiel (2012: 263), "[r]esponsibility is not a purely agent centred concept", rather it is "the art of the possible, sincerely harnessed in the benefit of care". Responsibility is associated with the ways in which dependencies and interactions are mediated, on terms that not only protect and provide for the human right to life, but fully enact responsibility in the protection and caring for all life – configured here as the co-presences that culminate in a culturally distinct place world. Instead of the human right to freedom being paramount, there is an overriding responsibility to support the conditions by which all-else experiences its freedom as a form of internal order and character. Rose (2008:110) describes these "relations of interdependence", as bearing responsibilities for others. This invokes a higher-order witnessing of harmful events, which may reconfigure violence in place, disorder and the greater effects of cultural wounding as a bodily experience intended to disrupt a person's physical and emotional sense of well-being, so too it may disrupt the epistemologies, ontologies and axiologies that support witnessing as a form of spectatorship, pressing harder than ever before into all human life, irrespective of closeness or distance from wounded spaces.

Emphasising responsibility along with rights discourse is about enhancing relational awareness and recognition as to the life worlds in which humans exist. I also contend that rights already invoke a discourse of responsibility, but distinguish this as exclusively human. The establishment of responsibility returns this discussion to a wider realm of kinship. Human roles and responsibility are derived of relational ontologies and awareness and it is kinship that structures cycles of responsibility (Pierotti 2008: 185). This is kinship as something pervasive, as more than people's biological and social relationships to one another, and inclusive of all and any elements of the place world in which humans are one element. A kincentric ecology functions in accordance with multifarious agents that extend beyond human life, thus enlarging our perceptual selves and the capacity to see other agents and presences of consequence. Kincentricity compels the awareness that other agents and co-presences possess and demand rights through their inherent character and order. With increased awareness, human responsibility is expanded into a field of greater relations where realising and investing in place order are the conditions of a responsive reflexivity.

Accenting human responsibility over human rights is not to de-emphasise the experience of disproportionate human suffering, nor dispute people's claims to enduring hardship and cultural wounding as breaches of their human rights. Instead, it is to locate the effects of such wounding and its lasting effects within a wider context of relations and, in doing so, highlight the relationship between human suffering and other localised suffering. This works to deepen the recognition of cultural wounding by noting the extent to which it might saturate into an entire 'whole world of life' within which individuals and cultural collectives are nested. If people lose their homes, are expelled from their territories or find their ancestral lands and waters poisoned by incoming agents, then the fate of those people is compounded by the fate of their places. Their struggle to prevail and rally around a cultural identity that has become untethered or lost within a disordered place world can therefore be more fully apprehended. This puts to rest any expectation that there might be a 'solution' to ethnic conflict and prevailing cultural wounding, which focuses solely on human rights, without serious consideration of human responsibility in the face of greater harms. It is the saturating effects of cultural wounding that imprint so hard on the place world that make so probable the cross-generational impact of trauma and violence. By acknowledging the depth of the wounds, for all that are nested in the receipt of harm, then healing becomes a multi-sited commitment in which the rights and freedoms for all must be realised.

Relational dynamics, uncertainty and rapid change

Testimonies of violence and trauma are given in families and communities, trials and truth commissions, newspapers, films and even in solitude (Weine 2006: xiii). Through spoken, written and visual languages, people tell stories that may change history, politics and life itself (Weine 2006: xiii). Testimonies of cultural wounding, violence and harm demonstrate the power of speech acts and strive to re-establish the individuality of victims and the collective identity of those who have lived through disordering events. These are the accounts to which many have become accustomed, and as testimonies they are powerful because they embody and express hope or keep at bay fears of disappearance by retrieving and sharing silenced memories of hardship and enduring suffering (Weine 2006). Late modern life, with its verbose tendencies, has been inclined to 'hear' certain testimonies over others and thus people have become used to the written or spoken discourse of violence and its wounding effects on human groups (Connerton 2011: ix). This has heavily shaped trauma studies and research on collective and individual human suffering and survival. While there is some recognition of a pre-linguistic human capacity to generate testimony, much less has been written on other types of testimonies, particularly those proffered by non-human agents; that is, the testimonial capacity of place and place elements.

Pre-linguistic testimonies reflect the situation in which lived bodies undergo events and then in turn retain and express (without words) the specificity of experience. This may be performed through bodily practices and postures as documents and texts (Connerton 2011: ix). Or it may be that memory is inscribed into human bodies (through, for example, genetic traits that map experiences of trauma and hardship, bodily habits, tattoos, scars and even styles of dress) and incorporated into institutions, histories and traditions (both as patterns in the built environment, educational bodies, governmental structures, and epistemological distinctions). This is, as Connerton (2011: ix) reflects, the moment when memory 'takes place' on and around the body's surface, and in its tissues, and in accordance with levels of meaning that reflect human sensory capacities more than cognitive categories. How the body becomes a witness to and expression of cultural wounding and its harmful effects is widely studied and examined through a discourse of healing and survivorship. If the pre-linguistic is accepted, and the non-verbal recognised as a fibre through which people remain connected to their experiences of cultural wounding, violence and harm, then by extension other non-verbal communicative events might also serve as testimonials to the effects of harm.

Conceiving of non-verbal testimonies liberates this discourse further, and realises the place world as capable of conveying the effects of wounding, uncertainty and rapid change.

As relational dynamics in place break down or struggle under the weight of uncertainty and rapid change brought on by the greater effects of cultural wounding, then testimonies of disorder and loss begin to emerge from place. What these testimonies might look like and how they might sound has been explored throughout the diagnostic Chapters 3, 4 and 5. They take the narrative form of trauma claims, etched in a language that may evade human translation where kincentric orientations are muted. Yet the disorder that is expressed as absence, silence, species decline, melancholia, spectral traces, human sickness and toxicity speak loudly of interruption in the relational dynamics that allow systems to prosper, survive or maintain integrity in particular structural and functional properties. Where disturbances compromise self-restoration in place, the testimony is often one of decline, disarray and loss of elements important to the whole. It may be as Linda Hogan, Chickasaw intellectual, academic, and poet writes:

> the earth speaks its symptoms to us. With the nuclear reactor accident in Chernobyl, Russia, it was not the authorities who told us that the accident had taken place. It was the wind. The wind told the story. It carried a tale of splitting, of atomic fission, to other countries and revealed the truth of the situation. The wind is a prophet, a scientist, a talker.
>
> (Hogan 2000: 118)

In writing of the Chickasaw corn harvest, her evocative words deliver the fullness of place and place elements to communicate needs and longing to a witnessing human kin:

> The leaves of the corn want good earth. The earth wants peace. The birds who eat the corn do not want poison. Nothing wants to suffer. The wind does not want to carry the stories of death.
>
> At night, in the cornfields, when there is no more mask of daylight, you hear the plants talking among themselves. The wind passes through. It's all there, the languages, the voices of wind, dove, corn, stones. The language of life won't be silenced.
>
> (Hogan 2000: 122)

The wounded place offers its testimony across all spaces and times. These narratives are known to exist, they are pre-linguistic and non-verbal iterations that communicate through mediums that often require culturally attuned awareness to be heard, seen, felt and understood. Scientific narrations of the world's physical decline, as biological, atmospheric and geological shifts are familiar, as are Indigenous accounts of ancestral suffering and the death of sentient beings in place, and the layperson's accounts of sadness, 'bad vibes' and overwhelming feelings of awfulness in place. How we come to listen and apprehend the testimonies of violence and wounding that are given in and by place depends on the cultural apparatus to which humans avail themselves. According to a kincentric principle, humans are co-presences and necessary witnesses to place. Beyond this, and because of this, we are compelled, through responsibility, to consider how rapid change and uncertainty, as it might compromise integrity and survival, impact upon the place world. Human life is not immune to the effects of rapid change and uncertainty in the place world, being both the agent of harm and the co-recipient of the hardships and suffering these experiences bring about.

There is not a singular vision of how testimonies of wounding, harm, uncertainty and change might be heard and deciphered for meaning, yet this book campaigns for certain cultural dispositions to be more widely shared and accepted into a normative practice of place relations. These are dispositions that make possible a responsive reflexivity to relational dynamics and that more fully recognise the struggle faced in periods of rapid and significant change. It is hoped that an increased awareness of and sensitivity to kincentric place relations might expand the perceptual capacity of broader human life to configure place as a sentient co-presence, to which we are but one relational element, albeit a very important one. A shift of this kind, in perceptual capacity, invokes the willingness to 'be with another in a manner which shifts the self', to 'practice a critical intimacy as an ethical way of coming to know something' and 'witnessing the other in its axiological richness'. In the first case, being with another (in this case place) extends our own presence beyond just that of human life and human rights. This is intended to disrupt a person's physical and emotional sense of individual or human-centric well-being. In doing so it also disrupts the epistemologies, ontologies and axiologies that support witnessing of place harm as a form of mere spectatorship. This is echoed through the principle of critical intimacy, which involves a willingness to open ourselves to place testimony where rapid change and uncertainty have taken hold or threaten to do so. As an ethical way of coming to know, humans adopt a responsive reflexivity that recursively shifts who we are and how we might live. Lastly, this radically expands the ability to recognise elements of great worth and value in the place world.

These three steps, which take us deeper into a relational encounter with place and unsettle human rights through the knowledge of rapid and harmful change, underlie Haraway's (2003: 50) point, whereby she announces "ethical relating within or between species, is knit from the silk-strong thread of ongoing alertness to otherness-in-relation". It is held that practicing critical intimacy with place and place elements as an ethical way of coming to know the world and witnessing place in its axiological richness above and beyond human utility are key to resilience in the face of cultural and environmental wounding. There is a restorative element to kincentricity, where people and place have the potential to more fully witness one another and perhaps usher in episodes of healing and a return to integrity, however defined. This leads to the question, do people thrive best when the relational dynamics of people and place are sustaining and ordered by a restorative principle, based in kinship and practice? And if so, can the power to heal be found in local empiricism and place order?

Local empiricism and place order

Place is distinguished by local empiricism and order. It is through incremental decline or the rapid assault of this order that it is prevented from enacting its own agency; its liberties in rhythm are taken away by agents of harm. The local empiricism that gives rise to place is expressed through induction, causation and causal explanation, as grounded in place itself. Induction – as the processes or actions that bring about or give rise to place – is multifaceted, and may have tangible or intangible qualities, as processes of physical becoming or ancestral enlivening shape and inspire the place world. The induction of place order is an expression of how the world comes to be as it is, and its ability to sustain and thrive is embroiled in relations of cause and effect, into which human life is recruited as kin, relational other, co-presence or agent of harm; the new species of trouble. The nature and effect of these relations are powerfully expressed through communicative events in place, which articulate not only the character of place but also its biography as populated by multifarious co-presences.

According to kincentric ecology, human life is one part of place's local empiricism, playing a role in the experience, rhythm and distinctiveness of place. As co-presences and kin, this role should maintain the integrity of place order, or else risk living with or dying from the consequences. When it disregards such conditions, and defiles instead, then the effects of harm are evident and bear down on all that is present within the place world, human and non-human. If, however, local empiricism is upheld, then integrity, defined as both a form of coherence, holism and durability, converges into an order of resilience. Resilience as a capacity to heal, or adapt in the face of adversity, when bound to a kincentric view of life, requires that people and place heal simultaneously, sharing in the recuperative essence of axiological return. The linking of resilience with kinship is echoed by Skolimowski's (1993: 7) living in 'right relations'. As an extension of this integrity, an axiological return to place is to be in a critical intimacy with place, it is to live in reverence, or "empathy fused with reverence" (Skolimowski 1993: 7). "Living with reverence on the earth is to watch, notice and live in heightened contact" (Skolimowski 1993: 7, cited in Cox and Holmes 2000: 73).

A vast body of literature speaks to the possibility of healing in the aftermath of violence, rapid change and wounding among ethnic groups (see Kearney 2014). Part of what motivates these final thoughts is the matter of whether the same can be said for place. It would seem that places can return to a state of integrity, they can be rehabilitated by human intervention, or become something new altogether without human involvement, as seen in the case of former villages that become thriving habitats for non-human species of fauna and flora. New places may also emerge in the aftermath of others declining, as both reflective of a human need for places of importance but also of the possibility of place genesis at any moment in time. There is vitality to the place world that is part of its local empiricism, perhaps this is even an ahistorical tendency that reveals the ontology of place as being one of movement, not fixity (see Merlan 1998: 215; Myers 1986). This quality of movement, however, does not extend so far that all forms of rapid change, uncertainty and violence can be accommodated. Some may go too far, as witnessed across the coastal stretches of the Yanyuwa people's beloved homelands, as discussed in Chapter 2. It may be that just as human life cannot always survive amid the violence of cultural wounding and trauma, some places also will not prevail.

What is it then that increases the likelihood of survival and resilience? Throughout this book, a commitment has been made to locating people and place along an equal register of vulnerability in moments of violence and cultural wounding. In essence, the experience is shared. So too is healing a shared business. It is therefore proposed that people thrive best when their place kin are strong, and that place thrives when its human counterparts are strengthened and valued for their cultural and ethnic distinctiveness. The ameliorating or even therapeutic effects of recognising Indigenous land rights in a settler colony are evidence of this, as are the accounts of populations or families returning home in the aftermath of expulsion and exile, finding a space in which to once again grow as a collective in strength and esteem. For those who find their belonging in diasporic spaces such as the pervasive and benevolent presence of an 'everywhere, and right here' Mama Africa, evidence suggests that healing is found in a return to place or even in the memory of place. Either way, the power of local empiricism and place is such that human life responds directly to its presence and intervention. When nested well within the place world, human life can flourish. And as for place, when nested alongside a reflexively responsive human presence, the likelihood of place order being recognised and safeguarded for its axiological worth is vastly increased.

Final thoughts

> Once I ground corn with a smooth, round stone on an ancient sloping metate. Leaning over, kneeling on the ground, grinding the blue corn, seeing how the broken dry kernels turned soft, to fine meal, I saw a history in that yield, a deep knowing of where our lives come from, all the way back to the starch and sugar of corn.
>
> (Hogan 2000: 122)

A book of this sort does not lend itself to conclusions. At best it can offer pathways that might assist in an axiological return, as a practice that brings human life into a more profound relationship with place. This is walked by way of kincentric ecology. Perhaps, as Hogan (2000: 122) writes, we begin this by looking upon place with a deep knowing that human life is contained in place, or is itself a part of place. It is born of local empiricism and is subject to place order. Denying this has not served human life well, yet there are those cultures that remain closer to this ontology, and seek to keep open the pathways of reverence that strive to safeguard place, to ensure its health and well-being and listen carefully when it communicates its ills. Indigenous epistemologies offer sophisticated models for an axiological return. Hogan explains:

> What we are really searching for is a language that heals this relationship, one that takes the side of the amazing and fragile life on our life-giving earth ... a language that takes hold of the mystery of what's around us and offers it back to us, full of awe and wonder. It is a language of creation, of divine fire, a language that goes beyond the strict borders of scientific inquiry and right into the heart of the mystery itself.
>
> (Hogan 2000: 121)

Written from the perspective of a witness, by someone who has seen the harms done to place and its people through an ethnographic lens and through responsive reflexivity, this book strives to find a new language. Having written elsewhere on cultural wounding and its effects on the lives of Indigenous groups in Australia and African descendants in Brazil, I find it unfathomable that the experience of violence and harm is read overwhelmingly in relation to human suffering. These forms of violence require not only a setting, but also a whole context of life to take hold and wreak havoc. I have seen people and place oppressed by the weight of abandonment, annihilation, disordering and toxicity, all of which betray a profound denial of kinship and axiological worth by those incoming agents who deliver harm. Fieldwork and ethnography has taught me that those who survive cultural wounding and its violent effects come through this process as something new, yet with powerful links to an overarching order of who they once were and who they aspire to be into the future. This desire for integrity in moments of uncertainty and rapid change is what unites human, non-human, place and place elements in the experience of wounding and violence. The place world, like its human kin, may survive, yet evidence shows that some violence might be so great that survival is not assured. In this case, place enters into the realm of memory, dying a death that is mourned heavily by human kin and a whole suite of kincentric place elements. Ultimately, the crisis lies in separating human life out from place and the solution lies in finding pathways back to kinship and relational dynamics of care and concern.

In describing the writing enterprise, American novelist Bob Shacochis (2001: 14–16), has shared with readers why it is that he writes. "I write for revenge" (Shacochis 2001: 3,15) he declares, a revenge that he describes as,

> Revenge against apathy, against those who are not interested in listening to the voices that surround them – wife, husband, brother, daughter, father, friend or nameless traveller.
> Revenge against the bullets of assassins, against the wild forces that trample the earth, against the terror and tragedy that is in every life.
> (Shacochis 2001: 14–16, in Soyini Madison 2005: 181)

Like Shacochis, I have written this book for revenge, revenge against the harmful agents that enact violence against the place world, as a consequence of violence against ethnic minorities and marginalised populations. It is revenge against the apathy and unresponsive reflexivity that characterises the disposition of those who normalise violence in place and turn away from the greater effects of cultural wounding. This type of revenge is productive in its desire to find epistemologies and ontologies that will support an axiological return to place value and kinship. This is found in a heightened awareness and commitment to human responsibility over human rights, realisation of the relational dynamics that embed and nest human life in the place world in an indivisible way, and recognition of the local empiricism and place order that gives each place and in turn its people their character and distinctive qualities. These steps invoke kincentricity, which sets the pace for another shift, that is the finding of language, or "a tongue that speaks with reverence of life, searching for an ecology of mind" (Hogan 2000: 122). If this step is not made, then the risk is that, "without it we have no home, have no place of our own within the creation".

> It is not only the vocabulary of science that we desire. We also want a language of a different yield. A yield rich as the harvests of the earth, a yield that returns us to our own sacredness, to a self-love and respect that will carry out to others.
> (Hogan 2000: 122)

References

Bhattacharyya, J. and Slocombe, S. 2014. *Indigenous Knowledge and Kincentric Ecology: Implications for Wildlife Management.* Available at https://warnercnr. colostate.edu/docs/hdnr/hdfw/2014/Presentations/063.M3.traditional_ecological_ knowledge/063.M3.001.indigenous_knowledge_and_kincentric_ecology_implic/ Pathways_Oct2014_Bhattacharyya_Slocombe.pdf. Accessed 7 May 2016.

Bradley, J. with Yanyuwa Families. 2016. *Wuka nya-nganunga li-Yanyuwa li-Anthawirriyarra: Language for Us, the Yanyuwa Saltwater People: A Yanyuwa Encyclopedic Dictionary.* Melbourne: Australian Scholarly Publishing.

Connerton, P. 2011. *The Spirit of Mourning: History, Memory and the Body.* Cambridge: Cambridge University Press.

Cox, H. and Holmes, C. 2000. Loss, Healing and the Power of Place. *Human Studies* Vol. 23: 63–78.

Finlay, L. 2003. The Reflexive Journey: Mapping Multiple Routes and Deconstructing Reflexivity, in *Reflexivity: A Practical Guide for Researchers in the Health and Social Sciences.* Edited by L. Finlay and B. Gough, pp. 3–20. Blackwell Publishing: Oxford.

Gough, B. 2003. Deconstructing Reflexivity, in *Reflexivity: A Practical Guide for Researchers in the Health and Social Sciences.* Edited by L. Finlay and B. Gough, pp. 21–36. Blackwell Publishing: Oxford.

Haraway, D. 2003. *The Companion Species Manifesto: Dogs, People, and Significant Otherness.* Chicago: Prickly Paradigm Press.

Hogan, L. 2000. A Different Yield, in *Reclaiming Indigenous Voice and Vision.* Edited by M. Battiste, pp. 115–123. Vancouver, British Columbia: UBC Press.

Kearney, A. 2014. *Cultural Wounding, Healing and Emerging Ethnicities: What Happens When the Wounded Survive?* New York: Palgrave Macmillan.

Merlan, F. 1998. *Caging the Rainbow: Places, Politics, and Aborigines in a North Australian Town.* University of Hawaii Press: Honolulu Hawaii.

Myers, F. 1986. *Pintupi Country, Pintupi Self: Sentiment, Place, and Politics Among Western Desert Aborigines.* Berkeley, California: University of California Press.

Pierotti, R. 2008. *Indigenous Knowledge, Ecology, and Evolutionary Biology.* Lincoln: University of Nebraska Press.

Regan, T. 2004. *The Case for Animal Rights.* Berkeley, California: University of California Press.

Rose, D.B. 2008. Dreaming Ecology: Beyond the Between. *Religion and Literature* Vol. 40(1): 109–122.

Salmón, E. 2000 Kincentric Ecology: Indigenous Perceptions of the Human-Nature Relationship. *Ecological Applications* Vol. 10(5): 1327–1332.

Scheper-Hughes, N. 1995. The Primacy of the Ethical: Propositions for a Militant Anthropology. *Current Anthropology* Vol. 36(3): 409–440.

Shacochis, B. 2001. Writing for Revenge, in *The Spirit of Writing: Classic and Contemporary Essays Celebrating the Writing Life.* Edited by M.R. Waldman, pp. 14–16. New York: Penguin.

Singer, P. 2011. *Practical Ethics.* Cambridge: Cambridge University Press.

Skolimowski, H. 1993. *A Sacred Place to Dwell: Living with Reverence Upon the Earth.* Rockport Massachusetts: Element.

Soyini Madison, D. 2005. *Critical Ethnography: Methods, Ethics and Performance.* California: Sage.

Weine, S. 2006. *Testimony after Catastrophe: Narrating the Traumas of Political Violence.* Evanston, Illinois: Northwestern University Press.

Yechiel M.B. 2012. *Human Dignity, Human Rights, and Responsibility: The New Language of Global Bioethics and Biolaw.* Cambridge: MIT Press.

Index

Abufarha, N. 105–6, 107
African Brazilians 16, 17–18, 19, 108, 144–6
African Diaspora 18, 144, 146, 147
agency 3, 8, 16, 20–1, 22, 29, 32, 34, 40, 193
agents of harm 3, 9, 86, 88–9, 95, 97, 98–104, 116–17, 174–5, 178–9, 182
ancestors 9, 19, 21, 22
ancestral beings 9, 22, 29, 51, 52, 55, 57, 79, 101, 102, 103
Anderson, B. 35
annihilation 4, 39, 72, 86, 104, 111, 204
Australia 36–7, 86; Indigenous homelands 9, 18, 29, 88, 129–30, 155, 172–4; non-human animals 180, 181; nuclear testing 172, 173; place names 55, 126, 127; see also Yanyuwa Indigenous country
axiological crises 3, 69, 71, 103–4, 112, 141–2
axiology 4, 68, 100, 103–4, 196

Battiste, M. 25, 26, 27
Belcourt, B. 181
Bhattacharyya, J. 194
Bikini Atoll 161, 167, 168–70, 171
Bing Bong 51, 52, 55–6, 62–8, 69, 71–2, 74–5, 76–7, 79, 80–5, 86–8
Borroloola 73, 74, 77–8, 79
Braverman, I. 106, 107
Brazil 16, 17, 18, 19, 108, 111–12, 144–6
bushfires 36–7

Cajete, G. 26, 27
Carpentaria, Gulf of 51, 52, 53, 55, 77
Cheslatta T'En 109–10, 112
colonialism 19, 53–4, 180–1
conflict 2, 5, 7, 8, 15

Connerton, P. 139, 142, 199
contamination 9, 81, 153, 156–7, 160, 161–2, 163–5; nuclear testing 166, 167–8, 169–71, 172–3; pollution 153, 156–7, 163–5
co-presences 3, 5, 16, 57, 102–3, 148, 182, 193, 202
Coward, M. 116
Cox, H.M. 36, 37
cultural wounding 1–6, 7–8, 15–16, 17–19, 34–5, 39, 89, 182, 183, 198–9; Bing Bong 52, 88
Cyprus 131, 132–8

dark tourism 7
Davis, D. 113
Davis, J. 168, 169, 170, 171
decolonising methodologies 9, 13, 16, 21, 22, 24, 158, 193
designification 17, 104–6, 107, 112, 117
destruction 17, 36–7, 38, 104–5, 106, 107–8, 109–12, 117, 174
disordering events 74, 98–101, 102–3, 124, 138, 140
displacement 36, 37, 109, 113, 115
Downing, T. 113
Dwyer, P. 29

ecocide 85, 86, 156–7
ecological decline 9, 153–4, 156, 163, 182
ecology 1, 7, 35, 86, 148, 153, 157–8
elemental erasure 9, 111, 148, 153–4, 156–7, 159–61
emotional geography 7, 16, 19, 37, 38, 39, 68–9, 108, 125–6
Enewetak Atoll 167–8
environmental damage 81, 83, 84–5, 109–10, 156–7, 163–5
Erikson, K. 3, 9, 98, 160–1, 170

ethnic cleansing 106, 113–14, 115, 142–3
ethnic conflict 7, 15, 17, 153, 174,
    175, 179–80
ethnic groups 7, 10, 17, 21, 124–5, 136,
    138–9, 141–4, 147, 153–4, 156
ethnic violence 8, 99, 132, 136, 142–3,
    144, 147

Falah, G. 105, 106
forgetting 139, 140, 141, 143–4, 147
Fullilove, M.T. 36, 37

geography 6, 29, 35, 156
Glencore 77–9, 80–3, 85
Goodall, H. 172, 173
Greek Cypriots 108, 132, 133,
    134, 135–6
Green Line, Cyprus 134–5

Halilovich, H. 142–3
harm 1, 2, 4, 7, 8, 9–10, 13, 21, 34–5,
    88–9, 199; agents of 3, 9, 86, 88–9, 95,
    97, 98–104, 116–17, 174–5, 178–9, 182
harmful places 125, 141, 144
healing 1, 2, 7, 8, 10, 17, 41, 183,
    191, 202–3
Heider, F. 112, 153
Hewitt, K. 4, 5, 105, 111
Hitchcock, R. 165
Hogan, L. 200, 204
Hokari, M. 23, 24
Holmes, C.A. 36, 37
human action 1, 35–6, 41, 103, 114,
    116–17, 197
human conflict 1, 2, 8, 15, 17, 174–5
human life 1–5, 7–8, 19, 20, 27–9, 31–2,
    34, 182–3, 201, 203
human rights 197, 198, 201, 205
hyper-relativism 3, 4

Indigenous Australians 17, 18, 19, 23,
    25, 28, 58–9, 73, 78, 100–2, 103, 108,
    154–5; *see also* Yanyuwa Indigenous
    community
Indigenous Brazilians 111–12
Indigenous epistemologies 3, 8–9, 13,
    16, 19, 21–4, 25, 27, 30, 40, 41, 162,
    194, 204
Indigenous homelands 15, 16, 17, 18;
    Australian 9, 18, 29, 88, 129–30, 155,
    172–4; *see also* Yanyuwa Indigenous
    country
Indigenous knowledge 21, 22, 24, 25–6,
    39, 100–2

Indigenous peoples 21, 24, 26, 27–30,
    112, 128; kinship 11–12, 57, 58, 72, 80,
    85–6, 88, 157

Johnson, D.M. 28
Johnson, J.T. 21, 23, 24

kincentric ecology 1, 9, 10, 158–9, 171,
    191, 193, 194, 202
kincentricity 36, 39, 193, 194, 201–2, 205
kinship 3, 16, 18, 19–20, 22, 23, 26–30,
    40, 86, 103–4, 140–1, 155, 191, 198;
    Indigenous peoples 11–12, 57, 58, 72,
    80, 85–6, 88, 157
Kirsch, S. 164

Langton, M. 58
local empiricism 140, 157, 202, 203, 204
Lynch, K. 159, 182, 183

Malpas, J. 7, 28, 30, 31, 35
Mama Africa 18, 144, 145, 147, 203
Marshall Islands 160, 161, 166–71
Massey, D. 22, 31, 33, 34, 35, 36
McVey, J. 110
melancholia 137, 138, 140
MIM (Mount Isa Mine) 74, 75, 76–7
moral disengagement 6, 69, 88,
    165–6, 182
moral ecology 1, 193
Muir, C. 158

natural disasters 9, 97, 98–101,
    102–3, 104
Navaro-Yashin, Y. 132–3, 137, 140
nested ecology 39–40, 69, 80, 166, 182–3
nesting 38–9, 40, 41, 158, 182
Nixon, R. 162
non-human animals 8, 22–3, 85, 86, 157,
    174, 175, 176–7, 179–81, 182
non-human life 8, 16, 82, 153, 156–7,
    161, 174, 177–9, 180
nuclear testing 160, 161, 166, 167–8, 169,
    172, 173

Ojibwa people 155–6
Ok Tedi River pollution 163–5
old people 52, 57, 72
oration 58, 59–62, 69–70

Palestine 105–7, 114
Parsons, W. 165
people 20–1, 23, 30, 32, 41, 114, 193–4
perspectivism 3, 22, 23

phenomenological approach 19, 20, 30
place 1–8, 13, 16–18, 19–22, 23–4, 30–4, 40–1, 80, 114–15, 193–4
place creation 22, 32, 35–6, 125
place death 88, 156–7, 204
place flooding 109–11
place harm 2–3, 4, 8–10, 41–2, 49, 73, 77, 85, 95, 97, 104, 197
place identity 25–6, 32, 33, 72, 106, 108, 110, 124, 128, 136–7
place-meaning 23, 26, 31, 32
place names 51, 55–6, 125–7, 128–31
place order 202, 203, 204
place trauma 37, 38, 39, 95, 142–3
poisoning 82, 160, 161–2, 163–5
political violence 115–16, 125
pollution 153, 156–7, 163–5
Povinelli, E. 59–60, 154, 161, 162, 172
proxemics 114, 115

Redmond, A. 22, 29
relationality 3, 23, 32, 35
renaming 124, 125, 126, 129, 131, 136, 137, 140
resignification 17, 105
responsibility 195, 196, 197–8, 205
responsive reflexivity 195, 196
Ricard, M. 175
Roberts, J.T. 127, 128
Roberts, T. 53–4
root shock 37–8, 52, 104, 117, 171–2, 174
Rose, D.B. 18, 25–6, 29–30, 58–9, 87, 127–8, 158, 159, 174–5, 198; wounded space 9, 16, 127

Salmón, E. 158, 177, 178, 194
Seamon, D. 32–3
sentiency 3, 5, 8, 22, 40, 57, 59, 166, 193
Shkilnyk, A. 155, 156
sign reading 59, 100–2, 154
slavery 16, 17–18, 19, 144–6
Slocombe, S. 194
Smith, L.T. 24
social disorder 124–5, 128, 131, 138–9, 140, 147–8
space 33–4, 35–6, 40

spatial anxiety 31
Sullivan, P. 158
survival 5, 100, 107, 183, 194, 203, 204
Syria 107–8

testimonies 6, 147, 159, 199–201
toponymic distinctiveness 2, 55, 124, 125, 128
Totten, S. 165
toxicity 7, 9, 160–3, 164, 165–6, 171, 172
trauma 2, 6, 7, 15, 16, 17, 99, 199, 200; place trauma 37, 38, 39, 95, 142–3
traumascapes 7
Turkish Cypriots 132, 133, 134, 135–6, 137–8

unresponsive reflexivity 195–6
urbicide 115–16

Villagra, A. 176
violence 1, 2, 5–6, 7, 8, 15–16, 17, 97, 117, 199, 204
Virunga gorilla troop 176
Viveiros de Castro, E. 22, 177

war 1, 2, 4, 5, 174, 175, 179–80
Western Desert Resources (WDR) 83–4
Wilangarra 51, 52–3
Wimberley, E. 38, 40, 158
Windsor, J. 110
witnessing 1, 6, 9, 13, 15–16, 143, 196–7, 201
wounded space 7, 9, 16, 127–8, 147, 200–1
wounding 41, 97, 117, 204

Yanyuwa Indigenous community 28–9, 41, 49, 51, 52, 54–8, 59, 68–72, 73–9, 82–4, 85–8, 102; oration 59, 60–2, 69–70; place names 55–6, 130–1; *see also* Bing Bong
Yanyuwa Indigenous country 41, 52–5, 68, 69–70, 71, 72, 73, 75–7, 79–80, 82
Youngblood Henderson, J. 25, 26, 27

zoo animals 179–80

Printed and bound by CPI Group (UK) Ltd, Croydon, CR0 4YY

22/10/2024

01777628-0014